5th Engineering Science & Technology International Conference 2020

AA007544

Held online due to COVID-19

IOP Conference Series: Materials Science and Engineering
Volume 990

Padang, Indonesia
5 - 6 August 2020

ISBN: 978-1-7138-3302-4
ISSN: 1757-8981

Printed from e-media with permission by:

Curran Associates, Inc.
57 Morehouse Lane
Red Hook, NY 12571

Some format issues inherent in the e-media version may also appear in this print version.

This work is licensed under a Creative Commons Attribution 3.0 International Licence.
Licence details: http://creativecommons.org/licenses/by/3.0/.

No changes have been made to the content of these proceedings. There may be changes to pagination and minor adjustments for aesthetics.

Printed with permission by Curran Associates, Inc. (2021)

For permission requests, please contact the Institute of Physics at the address below.

Institute of Physics
Dirac House, Temple Back
Bristol BS1 6BE UK

Phone: 44 1 17 929 7481
Fax: 44 1 17 920 0979

techtracking@iop.org

Additional copies of this publication are available from:

Curran Associates, Inc.
57 Morehouse Lane
Red Hook, NY 12571 USA
Phone: 845-758-0400
Fax: 845-758-2633
Email: curran@proceedings.com
Web: www.proceedings.com

TABLE OF CONTENTS

CHEMICAL ENGINEERING

MECHANICAL ENGINEERING

CIVIL ENGINEERING

IOP Conf. Series: Materials Science and Engineering **990** (2020) 011001 doi:10.1088/1757-899X/990/1/011001

Speech by the chairman of the committee

The honorable Rector of Universitas Bung Hatta, Prof. Tafdil Husni and the leadership. Honorable keynote speaker, Prof. M Tominaga from Saga University, Prof. Kamaruzzaman Sopian from Universiti Kebangsaan Malaysia, Prof. Tresna Dermawan from Universitas Indonesia, Prof. Iwan Vanany from Institut Teknologi Sepuluh November Surabaya, Prof. Reni Desmiarti from Universitas Bung Hatta.

The conference was initiated on 2011 and 2013 by the Faculty of Industrial Technology of Universitas Bung Hatta under the modest name of National Conference of Science and Technology (Resatek). The conference then expands and changes its name into the 3rd Engineering Science and Technology International Conference on 2016 and 2018. The conference adopted the present name and expanded its scope to international. The conference has facilitated the communication between scientists across applied in engineering technology science.

The theme of this conference "Applied Technology for Sustainable Development" is consistent with the progressive development of applied engineering technology, which enables us to reveal and comprehend how technology works towards better life of the society around the world. Due to COVID-19 pandemic, the conference was held online via Video Conference/online on 9 September 2020.

We do believe that the theme would present challenges for all of the engineers to develop sustainable energy and industrial processes, materials, conceptual and the design processes. Further, we do hoped that the outcomes of the conference could lead academicians and practitioners in this field to create better-quality-of-life in the future, while at the same time considering the impact of various activities towards our environment.

Chair of the Organizing Committee: Dr. Burmawi, ST., M.Si.

Secretary: Dr. Maria Ulfah, S.T, M.T.

International Scientific Committee:

1. Prof. M Tominaga (Saga University)
2. Prof. Dr. Kamaruzzaman Sopian (Universiti Kebangsaan Malaysia)
3. Prof. Aris A. Syntetos (Cardiff University, United Kingdom)
4. Assoc. Prof. Dr. Ardiansyah Syahrom (Universiti Teknologi Malaysia)
5. Assoc. Prof. Radwan Dweiri (Al Balqa Applied University, Jordan)
6. Assoc. Prof. Dr. Oki Muraza (King Fadh University Saudi Arabia)
7. Prof. Dr. Tresna P. Soemardi (Universitas Indonesia)
8. Prof. Iwan Vanany, ST., MT., Ph.D. (Institut Teknologi Sepuluh November, Indonesia)

Content from this work may be used under the terms of the Creative Commons Attribution 3.0 licence. Any further distribution of this work must maintain attribution to the author(s) and the title of the work, journal citation and DOI.
Published under licence by IOP Publishing Ltd

IOP Conf. Series: Materials Science and Engineering **990** (2020) 011001 doi:10.1088/1757-899X/990/1/011001

Organizing Committee:

1. Prof. Dr. Hendra Suherman, MT (Universitas Bung Hatta)
2. Prof. Dr. Eng. Reni Desmiarti (Universitas Bung Hatta)
3. Dr. Hidayat, MT (Universitas Bung Hatta)

IOP Conf. Series: Materials Science and Engineering **990** (2020) 011002 doi:10.1088/1757-899X/990/1/011002

Peer review declaration

All papers published in this volume of *IOP Conference Series: Materials Science and Engineering* have been peer reviewed through processes administered by the Editors. Reviews were conducted by expert referees to the professional and scientific standards expected of a proceedings journal published by IOP Publishing.

- **Type of peer review:** Type of peer-review is Single-blind.

- **Conference submission management system:**
  Conference submission management system is use website. Website can be accessed at https://esticfti.bunghatta.ac.id/. Corresponding author process by e-mail ESTIC: estic2020@bunghatta.ac.id.

- **Number of submissions received:**
  Number of submissions received about 52 manuscripts.

- **Number of submissions sent for review:**
  Number of submissions sent for review is 48 manuscipts.

- **Number of submissions accepted:**

  Number of submissions accepted is 34 paper.

- **Acceptance Rate (Number of Submissions Accepted / Number of Submissions Received X 100):**

  Acceptance rate = 34/52 x 100 = 65.38

- **Average number of reviews per paper:**

  Average number of reviews per paper = 48/52 x 100 = 92.31

- **Total number of reviewers involved:**

  Total number of reviewers = 6.

- **Any additional info on review process:**
- **Contact person for queries:**

  1. Dr. Burmawi, S.T., M.Si. (+6282387464859, burmawi@bunghatta.ac.id)
  2. Maria Ulfah, S.T., M.T. (+6281321437068, mariaulfah@bunghatta.ac.id)
  3. Drs. Mulyanef, S.T., M.Sc. (+6281374005188, mulyanef@bunghatta.ac.id)
  4. Budi Sunaryo, S.T., M.T. (+6281320777313, budi.sunaryo@bunghatta.ac.id)

Content from this work may be used under the terms of the Creative Commons Attribution 3.0 licence. Any further distribution of this work must maintain attribution to the author(s) and the title of the work, journal citation and DOI.
Published under licence by IOP Publishing Ltd

IOP Conf. Series: Materials Science and Engineering **990** (2020) 012001 doi:10.1088/1757-899X/990/1/012001

Development of Machine Learning Applications for Chemical Objects

ASetiawan[1], S Rostianingsih[1], CI Halim[1]
[1]Informatics Department, Petra Christian University, Surabaya, Indonesia

E-mail: alexander@petra.ac.id

Abstract. Machine learning is a method used to create programs that can learn from data or applications. In contrast to ordinary static computer programs, machine learning programs are programs designed to be able to learn on their own. Machine learning technology has been widely applied in various fields, including in chemistry. Currently the introduction of basic chemical objects is a study that has many concepts that are easy for beginners to understand and imagine, such as atomic structure and molecular shape in chemical bonds. So that an alternative learning media is needed in order to improve understanding of the concept of chemical objects interactively. Machine learning technology is expected to provide a more realistic interaction and is an advancement of a promising technological method that can motivate users to engage in a more active learning system.

1. Introduction

The development of a technology is getting faster and faster. Technology and the development of machine learning is one of the most frequently developing fields of science today. Without realizing it, almost everyone uses products produced by machine learning applications. The term in machine learning is basically the computer's process of learning from data. All technology in machine learning will inevitably involve data. The data can be the same, but an algorithmic concept and a different approach to get the optimal level of ability and results. Presenting chemical processes for students with visualization is highly emphasized in chemistry education, because it can help students develop an understanding of chemical concepts to find the implications of using machine learning in chemistry learning, which is to encourage active, organized, and able learners to integrate the information obtained to understand the concepts and difficult principles and can be used to solve problems. Therefore, important steps that need to be taken are providing learning resources for students to overcome difficulties in understanding chemical concepts.

One technique for applying machine learning is supervised learning. As discussed earlier, machine learning without data will not work. Therefore, the first thing prepared is data. Data will usually be divided into 2 groups, namely training data and testing data. The training data will be used to train the algorithm to find a suitable model, while the testing data will be used to test and determine the performance of the model obtained at the testing stage.

Machine Learning is used to recognize a pattern as is done by our brains. Therefore, we need input, model, and output. When Machine Learning does a new job, it will see what work has previously been done so that it can provide the correct output. Google Cloud Platform has developed tools for users to

[1] alexander@petra.ac.id

Content from this work may be used under the terms of the Creative Commons Attribution 3.0 licence. Any further distribution of this work must maintain attribution to the author(s) and the title of the work, journal citation and DOI.
Published under licence by IOP Publishing Ltd

develop Machine Learning in applications. Therefore, to continue previous research, namely 3D Modelling of Chemical Elements with Augmented Reality [1], an introduction to chemical objects was made using machine learning. The system has stored a set of datasets containing chemical objects and their element names. When a new object is scanned with this application, it will be matched with the previous dataset. If it turns out to be recognized it will appear the name of the chemical element. If not recognized, a new dataset can be added based on the results of the object recognition training. The desired end result is to display the chemical element names for the recognized objects.

2. Machine Learning

In machine learning, pattern recognition or feature extraction starts from the initial set of data and builds features that are meant to be informative, facilitating learning and subsequent generalization of steps, not superfluous. For some other cases, machine learning leads to better human interpretations. Several features are closely related to dimensional reduction. When inputting data into an algorithm, the size may be too large to be processed and may be overkill, the measurements are the same in meters and feet. Image repetition is represented as pixels. These can be turned into a reduced feature set (known as a vector pattern). In this process, it is known as pattern extraction. It is hoped that the extracted pattern will contain a relevant information from the input data. Therefore, the desired result is that it can be used for scaled down representations, not the initial complete data [2].

All of this means it's possible to quickly and automatically generate models that can analyst larger, more complex data, and provide faster and more accurate results - even at very large scales. And by building the right model, organizations stand a better chance of identifying profitable opportunities - or avoiding unknown risks. Lots of machine learning the researchers performed their work on single-purpose and general-purpose computers [3,4], and multiple single-machine frameworks supported this scenario. Caffe [5] is a high performances framework for training that is described ina declarative manner neural networks on multicore and general purposes computer purposes. As discussed above, the programming model is similar to DistBelief, so it is easy to model from existing layers, but relatively difficult to add new layers or optimizers. Torch [6] offers a powerful imperative programming model for scientific computation and machine learning. This allows fine grained control over command execution and memory usage, allowing power users to optimize performance their program. While this flexibility is beneficial for research, Torch does have the advantages of graphical data flow as a portable representation across small-scale experiments, production training, and deployment [7].

3. Research Methodology

The research methodology of machine learning application is very important for the success. Therefore, machine learning applications at student must be designed and developed so that they can be used. Testing machine learning must be done, and to received feedback from participants so that the level of application. Synthetic chemistry methodology refers to the methods used for the synthetic dataset objects can be seen in Figure 1.

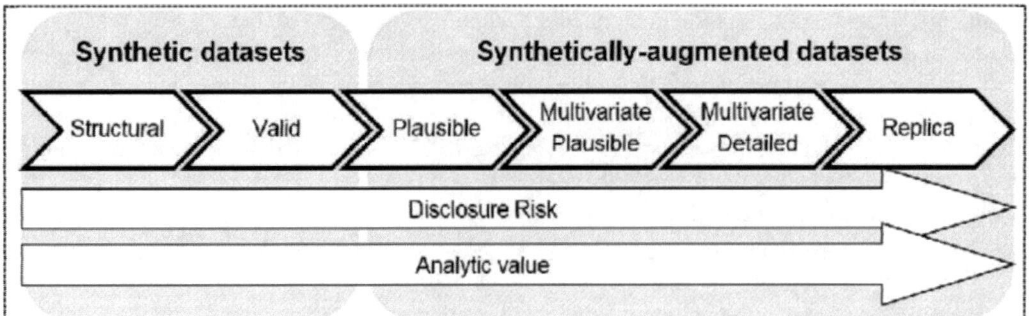

Figure 1. The synthetic datasets.

4. Implementation System

The implementation can be used Jupyter Notebook, Tensorflow, Python 3.6, CUDA Toolkit 10.0 and CUDNN 7.4.1. For machine learning development, Google Cloud Platform provides a machine learning API, while for image or voice recognition it uses TensorFlow. The programming language used is Python. Machine learning API is used to detect images (Cloud Vision API) such as logos or photos, video, natural language (text), voice, and translators. The five of them can be combined into one. From a set of datasets, machine learning (training, deployment, server) is carried out, then predictions are generated using the REST API.

The training dataset, you can use two ways, namely transfer learning or training from scratch. Transfer learning means updating existing data using and conducting training with your own data. While training from scratch is doing training from scratch with only your own data. After the training is complete, two hooks are tested: Testing of synthetic images / combined images, and testing with real images / real photos. Displays several sample images to confirm whether the neural network can read the image correctly can be seen Figure 2.

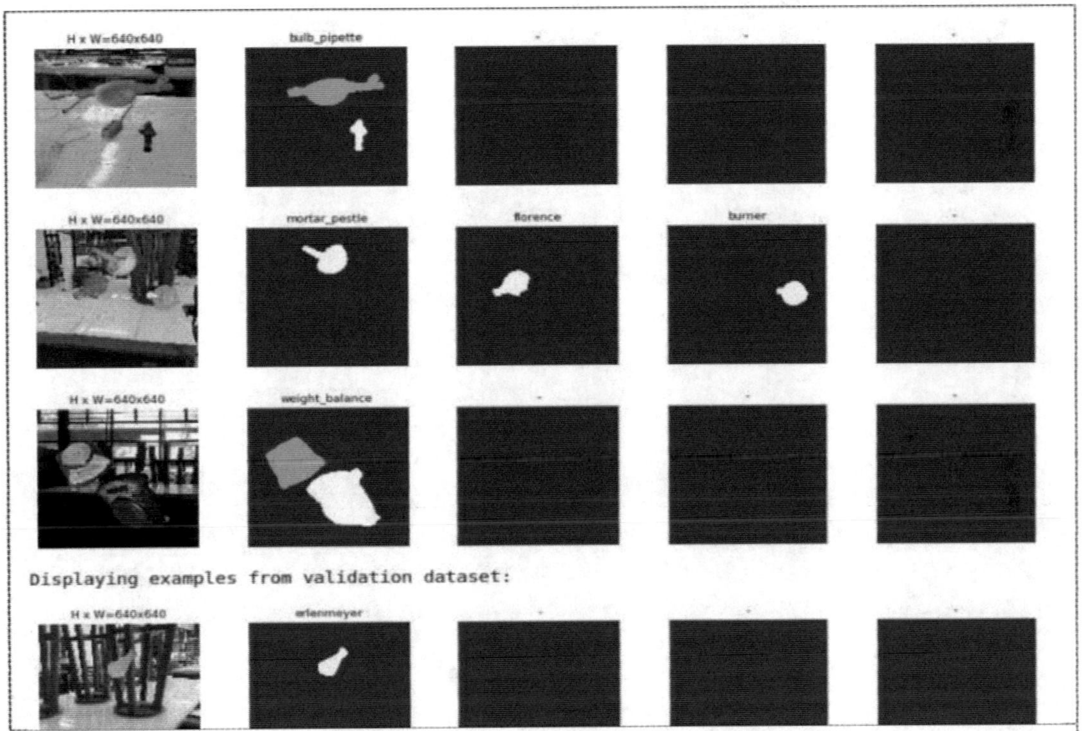

Figure 2. The testing of synthetic dataset.

Training Process:
Training: 4 Epochs, each 1000x cycles (total 4000 cycles)
Fine-tuning: 4 Epochs, each 1000x cycles (total 4000 cycles)
Total 8 epochs or 8000 cycle7.

The prototype that uses Google Machine Learning as the basis of this research, keep in mind that this neural network is being trained using a synthetic dataset, and according to result The Synthetic of Burner and Spektrofotometer can be seen Figure 3, The Synthetic of Weight Balance can be seen Figure 4, The Synthetic of Bulb Pipette can be seen Figure 5, and The Real Synthetic of Weight Balance can be seen Figure 6, when tested to recognize real photos, the results are quite satisfying, namely being able to recognize some of the objects/objects correctly.

IOP Conf. Series: Materials Science and Engineering **990** (2020) 012001 doi:10.1088/1757-899X/990/1/012001

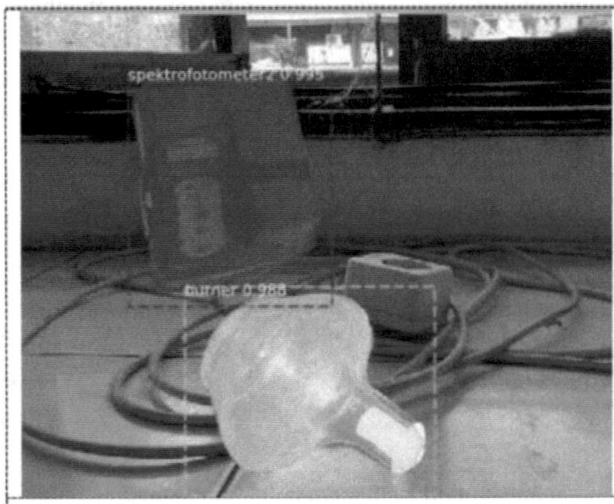

Figure 3. The synthetic of burner and spektrofotometer.

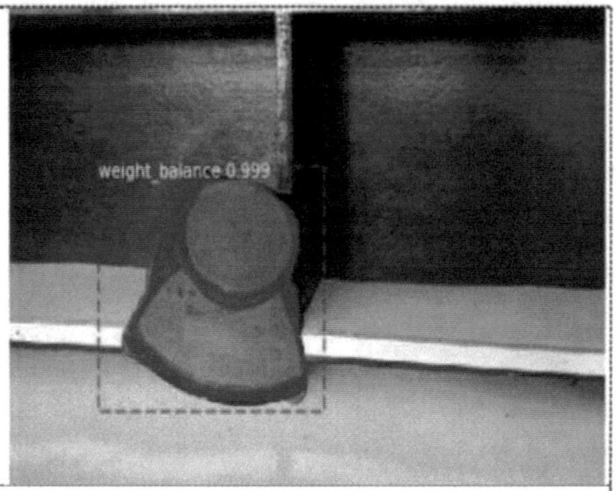

Figure 4. The synthetic of weight balance.

Figure 5. The synthetic of bulb pipette.

Figure 6. The real synthetic of weight balance.

5. Conclusion

The machine learning applications for chemical object have the problem is that most of the lab equipment is made of glass / transparent, such as: erlenmeyer flask, florence flask. It is more difficult for the neural network to recognize correctly, because the atmosphere / colour behind the glass objects affects the performance of the training data. The manual annotation process requires a lot of labour (human labour) and takes a long time to get meaningful results. Because machine learning requires at least thousands of unique images. Using synthetic datasets can speed up the creation of datasets, when training, can produce results that are not bad.

Acknowledgments

This Research Project was funded by a Internal Research Grant, received in 2020. We thank the Centre of Research Petra Christian University for the supports and guidance.

References

[1] Setiawan A, Rostianingsih S and Widodo T R 2018*MATEC Web Conf.* **248**05007

IOP Conf. Series: Materials Science and Engineering **990** (2020) 012001 doi:10.1088/1757-899X/990/1/012001

[2] Yadaf S K and Tiwari A K 2015*Mach. Learn. Appl.***2(3/4)**30-49
[3] Krizhevsky A 2014 One weird trick for parallelizing convolutional neural networks arXiv:1404.5997
[4] Krizhevsky A, Sutskever I and Hinton G E 2012 *Adv. Neural Inf. Process. Syst.* **25(2)**1097-1105
[5] JiaY,Shelhamer E, Donahue J, Karayev S, Long J, Girshick R, Guadarrama S and Darrel T 2014*Proc. 22nd ACM International Conference on Multimedia* 675-678
[6] Collobert L, Bengio S and Marithoz J2002*IDIAP Research Report*
[7] Abadi M et al. 2016*Proc. 12th USENIX Symposium on Operating Systems Design andImplementation* 265-283.

IOP Conf. Series: Materials Science and Engineering **990** (2020) 012002 doi:10.1088/1757-899X/990/1/012002

Development of Web-Based TOEFL Learning Media (E-TOEFL) Using Moodle at Universitas Bung Hatta

R Amelia[1]*, F Harmaini[2]
[1]Computer AndNetworkEngineering TechnologyDepartment, Faculty of Industrial Technology, Universitas Bung Hatta.
[2]English Education Department, Faculty of Teacher Training And Education, Universitas Bung Hatta

*riskaamelia@bunghatta.ac.id

Abstract. This research aims to produce a website-based TOEFL learning system (E-TOEFL) used for online TOEFL learning at Universitas Bung Hatta. The research method used in this research is a research and development (R&D) method. The steps taken in this study are the first stage that is to analyze the need to develop an E-TOEFL system-based learning management system (LMS), namely with observation and interview. Then carried out the design of this LMS-based E-TOEFL system. The next step is system testing by testing access speeds, link structures, and testing with a wide variety of browsers. The last step is the system implementation phase. In the implementation of E-TOEFL at Universitas Bung Hatta, some students have not participated in this E-TOEFL because TOEFL does design to be self-learning. So that in the future, it is necessary to apply blended learning to increase the enthusiasm of students in learning Toefl. More development also needs to be done to optimize online learning activities. One of the steps that need to do is to socialize continuously related to TOEFL's learning and simulation to students to make students more motivated to use this E-TOEFL. Then there is also the dissemination of E-TOEFL learning policies to the academic students of Universitas Bung Hatta..

1. Introduction

In today's globalization era, human resources' quality is key to competitiveness between countries. To participate in taking an active role inrelations between countries globally and especially the ASEAN region, required to engage in communication skills, especially in English. ASEAN Economic Community (AEC) is the door for ASEAN countries to cross into Indonesia. As a logical consequence, many foreign workers will visit and even find work in Indonesia. In this case, Indonesia must immediately prepare human resources that can speak English actively[1].

Aware of the importance of English language usage and its use in various fields, since 2013, Universitas Bung Hatta (UBH) conducted a TOEFL Like test with 400 as a student graduation requirement. However, when Universitas Bung Hattawas implementing this test, it was not as expected. From the TOEFL test result in February 2017, data from *Language Development Centre* obtained that all participants who took the TOEFL test as many as 1600 people, only 1,220 participants passed[2].

[1] riskaamelia@bunghatta.ac.id

Content from this work may be used under the terms of the Creative Commons Attribution 3.0 licence. Any further distribution of this work must maintain attribution to the author(s) and the title of the work, journal citation and DOI.
Published under licence by IOP Publishing Ltd

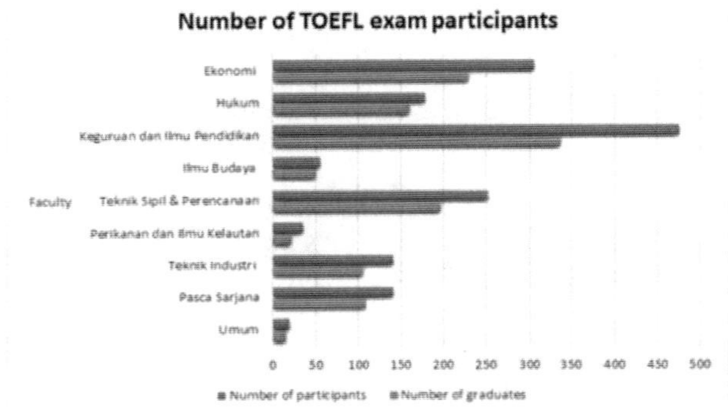

Figure 1. Number of participants who took the TOEFL test and participants who were declared to pass the TOEFL test

At the test time, Universitas Bung Hatta participants graduated with a min ≥ 400 score; people totaled 1,205 people (76.12 % of the total participants). Meanwhile, outside Universitas Bung Hatta, participants receive a ≥ 400 number of 15 people (78.94% of the full participants). Although participants from Universitas Bung Hatta who scored ≥ 400 reached 76.12 %, where this percentage is good enough, not all participants passed one test. Some participants had to repeat the test many times until they earned a score of ≥ 400[2].

As a result, students with fewer than 400 grades, many of them not graduating, have not passed the test despite having taken many exams. Moreover, this problem always occurs every semester. Shows that the low ability of students in English language skills, especially TOEFL. Based on interviews with several Private College stakeholders in Padang City, many of its students graduated TOEFL but with unspecified grades. Therefore, a solution is needed to solve the above problems by designing e-learning media. E-learning is a web-based learning ecosystem for information dissemination, communication, and knowledge for education and training that integrates multiple stakeholders with technology[3][4]. Research [5] proves that the use of e-learning can improve students' understanding. Besides, expected does e-learning make it easier for students to learn through e-learning and access it anytime and anywhere. Reflecting on the above issues, the design of TOEFL e-learning is crucial for students at UBH.

2. Literature Review

2.1 E-Learning.

The acceleration of the development of Information and Communication Technology (ICT) helps the learning process. Many educational institutions to adopt internet-based learning, which is widely known as e-learning or commonly LMS [6]. The increasing need for information has become a worldwide concern for utilizing new tools to exchange knowledge by fostering knowledge management and e-learning development, commonly referred to as the Learning Management System [7]. For many countries with large populations, limited resources and remote geographical areas such as Indonesia, E-learning or learning management systems (LMS) hold the key to better education and access to knowledge resources [8].

2.2 Type of Learning Management System.

Some examples of LMS are Moodle, Claroline, eFront, Schoology, ATutor, Sakai, Dokeos, LAMS, WebCT, and many others. Modular Object-Oriented Dynamic Learning Environment (Moodle) is part of an LMS (Learning Management System) that was introduced by Martin Dougiamas, a computer scientist and educator [9]. E-learning has the advantage of exceeding the limits of space and time, increasing comfort and effectiveness for individual and collaborative learning. Then providing information that can be reused and updated through the use of interactive multimedia [10] [11] [12] [13] [14]. According to [15], e-learning can increase student motivation at the university level's learning process.

3. Method

The research method or research design used is the research and development (R&D) method. The 4D research and development model has filled over four main stages, defining, designing, developing, and disseminate.

4. Result and discussion

The first step that researchers did wasconduct observations and interviews at the head of the foreign language development center of Universitas Bung Hatta. Based on the observations and interviews with the head of the center for foreign language development (PPBA) obtained data, many students have to repeat taking the TOEFL exam because they did not pass the TOEFL test. After being traced, it turns out that students do not understand what material to learn during the TOEFL test. In contrast, the material they get when studying discusses the basics of the English language only. The materials learned by students are not prepared to face the TOEFL test.

Furthermore, after conducting the activity, researchers began designing and developing an e-learning system as a simulation of English learning for students of Universitas Bung Hatta, which can then run effectively in improving the quality of education at Universitas Bung Hatta. The results of the design carried out in figure 2.

Figure 2. E-TOEFL homepage

In this E-TOEFL system, there are three users, namely: first admin as an application manager, who can manage the user and manage the course, the second is the lecturer who has the function to manage lecture materials, training, test, and assessment.

Figure 3. Display of TOEFL material menus

The type of TOEFL used in this system is the paper-based test type. The materials discussed in this TOEFL are Listening Comprehension, Written Structure and Expressions, and Reading Comprehension. In this E-TOEFL system, hundreds of question banks also allow students to practice and deepen the TOEFL material.

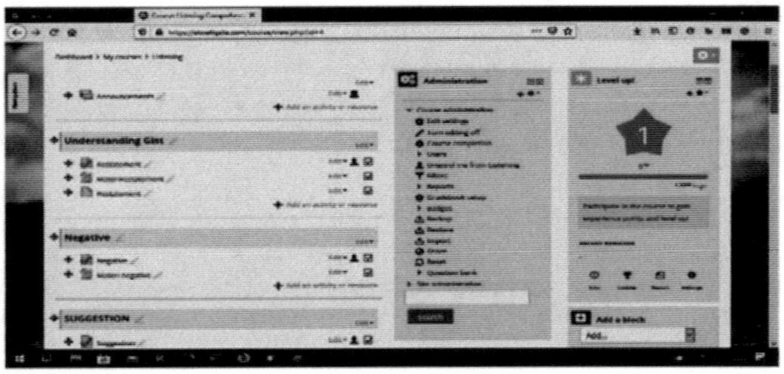

Figure 4 Display of the content of the listening material

Moreover, third, students who have the function to learn TOEFL materials uploaded by PPBA operators, practice, test, and see grades. To be able to access materials, forums, and exercises, users log in to the account that has been provided by the administrator. In E-TOEFL, there are various features, namely materials, exercises and quizzes. On the E-TOEFL system, game features do also provided. This form of gaming in this system is an exercise by guessing words in English. This game create intended to reduce the level of saturation of students in accessing e-learning materials.

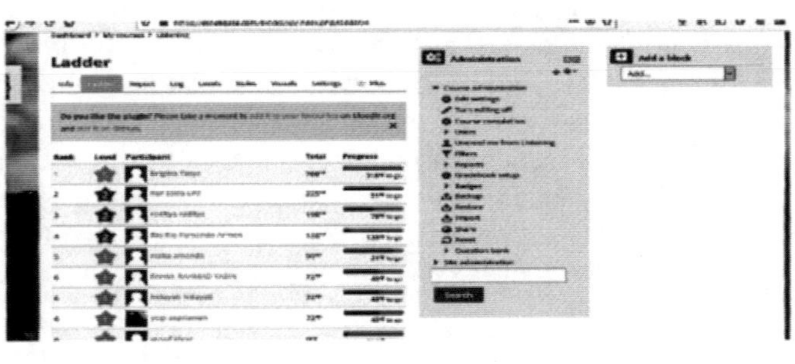

Figure 5. Display of the results of the score of the student who successfully worked on the exercise

To be able to access materials, forums, and exercises, users log in to the account that has been provided by the administrator. In E-TOEFL, there are various features, namely materials, exercises, and quizzes. On the E-TOEFL system, the game features do provide. The game form on this system is an exercise by guessing words in English. This game created intended to reduce the level of saturation of students in accessing e-learning materials.The third stage is the System Trial. Before the system is published, several tests do first carried out to ensure the system runs appropriately. The parts of the system that are tested include:

4.1 Access speed.
In measuring access speed on storage media or servers, testing was conducted on the Internet network using multiple web browsers. The results test showed that the system could run properly on any browser.

4.2 Link structure on the e-learning system.
Link structure needs to be tested to avoid possible links being disconnected due to the destination link, the change of the intended file names, the deletion of the destination link, or for other reasons. Testing is done by performing a one-on-one check of each menu connecting pages from one page to another. This test shows that the links contained in the system can run properly according to their respective functions.

4.3 Browser compatibility is used.
At this stage of testing browser compatibility against online learning applications, trials are conducted using several web browsers, including Internet Explorer, Mozilla Firefox, and Google Chrome. Based on the results of the trial, there was no problem with the E-TOEFL system trial.

4.4 Implementation.
The implementation of E-TOEFL at Universitas Bung Hatta does access to the website address www.etoeflgate.com. This E-TOEFL system developed using the moodle learning management system, which in this E-TOEFL gets 3 (three) user accounts, namely Admin, PPBA Operator, and Student.The use of moodle-based E-TOEFL as a simulation of TOEFL learning at Universitas Bung Hatta does expect to impact TOEFL test results. It can improve students' ability in English anywhere and anytime without being constrained by space and time..

5. Conclusion
That E-TOEFL application using moodle can do used in online TOEFL learning and simulation at Universitas Bung Hatta. Moodle has facilities that support the learning process. The moodle facilities need to be continuously explored and developed in online learning applications to support the improvement of the quality of learning systems at Universitas Bung Hatta, especially in TOEFL learning. In implementing E-TOEFL at Universitas Bung Hatta, some students have not participated in this E-TOEFL because TOEFL does design to be self-learning. So that in the future, it is necessary to apply blended learning to increase the enthusiasm of students in TOEFL learning. For further development on the utilization of LMS moodle-based e-learning as a learning medium at Universitas Bung Hatta. Optimize online learning activities; There needs to be ongoing socialization related to TOEFL learning and simulation to students to make students more motivated to use this E-TOEFL. Then there is also the dissemination of E-TOEFL learning policies to the academic students of Universitas Bung Hatta.

Acknowledgment
Researchers are part of the Higher Education funding research in 2019. Researchers say thanks to "DRPM Ministry of Research and Technology of Higher Education Contract No: 028-01 / LPPM-Penelitian / Hatta / IV-2020" that has funded this research.

References
[1] Shobikah, N. The Importance of English Language in Facing Asean Economic Community (AEC). At-Turats, 11(1). https://doi.org/10.24260/at-turats.v11i1.873. 2017
[2] Laporan TOEFL Wisuda 68. Agustus 2017. PusatPelatihan Bahasa Asing. *Unpublished*.
[3] B. Ghirardini, *E-learning methodologies: A guide for designing and developing e-learning courses*. 2011.

[4] W. A. Cidral, T. Oliveira, M. Di Felice, and M. Aparicio, "E-learning success determinants: Brazilian empirical study," Comput. Educ., vol. 122, no. July, pp. 273–290, 2018, doi: 10.1016/j.compedu.2017.12.001.

[5] E. O'Neill, N. T. Stevens, E. Clarke, P. Cox, B. O'Malley, and H. Humphreys, "Use of e-learning to enhance medical students' understanding and knowledge of healthcare-associated infection prevention and control," *J. Hosp. Infect.*, vol. 79, no. 4, pp. 368–370, 2011, doi:10.1016/j.jhin.2011.08.008.

[6] C.-H. Chung, L. A. Pasquini, and C. E. Koh, "Web-based learning management system considerations for higher education," *Learn. Perform. Q.*, vol. 1, no. 4, pp. 24–37,

[7] Yordanova, K. "Integration of Knowledge management and E-learning – common features" *International Conference On Computer Systems And Technologies* Rousse,Bulgaria. 2007

[8] Kuntoro, Ratna and Hawamdeh, Suliman.E-Learning in Higher Educational Institutions in Indonesia. *Journal Of Information & Knowledge Management. Vol*2, no. 4, pp361-374. 2003

[9] Fendi, Pipit Yusuf. Pengembangan*E-Learning* Berbasis*Moodle* UntukPesertaKursus Di Swift English School Yogyakarta. *E-Jurnal Prodi TeknologiPendidikan. 6(8). 201.*

[10] Yu S And Yang K. "*Attitudes Toward Web-Based Distance Learning Among Public Health Nurses In Taiwan: a Questionnaire Survey*". Int J Nurs Stud.; 43(6):767–774. 2006

[11] Choules Ap. *The Use OfElearning In Medical Education: a Review Of The Current Situation*. Postgrad Med J.; 83(978):212–216. 2007

[12] Cook Da. "*Web-Based Learning: Pros, Cons And Controversies". Clinical Medicine*. 7(1):37–42. 2007

[13] Moreira Ic, Ventura Sr, Ramos I, And Rodrigues Pp. *"Development And Assessment Of An e-Learning Course On Breast Imaging For Radiographers: a Stratified Randomized Controlled Trial"*. J Med Internet Res.; 17(1). 2015

[14] Al-Neklawy And Ahmed Farid. "*Online Embryology Teaching Using Learning Management Systems Appears To Be a Successful Additional Learning Tool Among Egyptian Medical Students*". Annals Of Anatomy. Http://Dx.Doi.Org/10.1016/j.Aanat. 2017.

[15] Safiyeh, RajaeeHarandi. (2015). "*Effects Of e-Learning On Students' Motivation*". International Conference On Leadership, Technology And Innovation Management. Procedia - Social And Behavioral Sciences, 181. 423 – 430.

IOP Conf. Series: Materials Science and Engineering **990** (2020) 012003 doi:10.1088/1757-899X/990/1/012003

WebGIS of Participants in the Webinar on Communication and Learning Motivation during the Covid-19 Pandemic

B Sunaryo[1], Hidayat[2], Zulfadli[3],A Yanuarafi[4],VNAnwar[5]
[1,3]Department of Network Computer Engineering Technology, Universitas Bung Hatta, Padang, West Sumatera, Indonesia
[2]Department of Electrical Engineering, Universitas Bung Hatta, Padang, West Sumatera, Indonesia
[4]Department of Informatics Engineering, Universitas Putra Indonesia YPTK, Padang, West Sumatera, Indonesia
[5]Department of Teacher Training of Education, Universitas Bung Hatta, Padang, West Sumatera, Indonesia

E-mail: budi.sunaryo@bunghatta.ac.id

Abstract. Effective communication is the key to the learning process for educators. Strong motivation is needed, especially during the Covid-19 pandemic. Webinars are a good medium for motivators to convey motivation and communication solutions for educators. To map the interests of the webinar participants, a Web Geographic Information System (WebGIS) was built. Online registration is spread through social media networks so that there are 859 participants spread across Indonesia. The first largest was obtained by West Sumatra Province with 21%, East Java Province 16.4%, and West Java Province 10%. Participants were dominated by teachers 48% and students 21.7%.With the presence of information on areas, numbers, and percentages visually through the WebGIS application, universities, institutions or stakeholders can formulate policies and planning programs to improve the quality of teaching and learning through communication and learning motivation, both for teachers and students.

1. Introduction

Communication in education is very important for learning achievement[1]. Students are required to be able to understand every material presented by the teacher[2], One of the conditions can be measured from the achievement of learning outcomes [3]. The Covid-19 pandemic has greatly affected the teaching and learning process in Indonesia[4]. The difference in learning methods becomes an obstacle for teachers and students in communicating[5][6]. Teachers and students need motivation that can increase enthusiasm in communicating their lessons, in order to create characters of honesty, tolerance, hard work, creative, independent, curious, communicative and responsible[7].

Various efforts have been made by universities and institutions in increasing teaching and learning motivationas Online Teaching-Learning[8][9]. One of them is by holding a webinar about communication and learning motivation during the Covid-19 pandemic. Universitas Bung Hatta, especially the Industrial Technology Faculty, took the initiative to hold the webinar activity.This

[1] budi.sunaryo@bunghatta.ac.id

Content from this work may be used under the terms of the Creative Commons Attribution 3.0 licence. Any further distribution of this work must maintain attribution to the author(s) and the title of the work, journal citation and DOI.
Published under licence by IOP Publishing Ltd

webinar was held on Sunday, July 12, 2020 with the speaker, Communication Expert, Dr. Aqua Dwipayana.

In general, participants who take part in this webinar consist of several academic elements from all over Indonesia. The participants are teachers, lecturers, students, and others. The distribution of their locations is recorded in the provinces in Indonesia. The enthusiasm of the participants in participating in this activity was still strong even though only through online media.

Based on the distribution of the participant's locations, a web-based Geographic Information System (GIS) application, known as WebGIS, was built. WebGIS is a computer-based information system that can retrieve geo-referenced data[10].The purpose of developing this WebGIS is to display and see the percentage of enthusiastic participants in the distribution map. This distribution data is the main dataset[11] needed in the MySQL database.

2. Methods

The research method focuses on the distribution of participant data locations. Starting from data collection, data analysis, system development, system testing, and evaluation, and conclusions. These stages are illustrated in Figure 1.

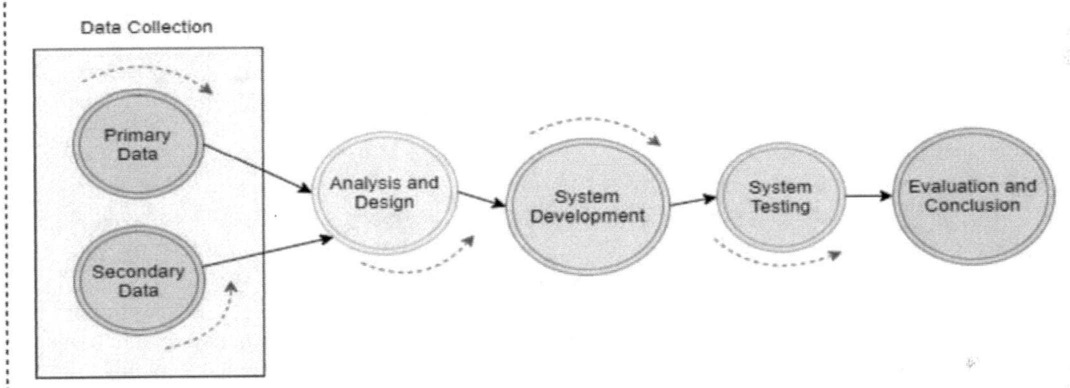

Figure 1. Research methods.

2.1. Data Collection

Primary data were obtained from the results of registration via Google Form as many as 859 participants. In Google Sheets, several data fields are taken, namely e-mail, full name, age, contact number, profession, agency, address, province, and motivation. The field that becomes the focus of research for mapping needs in WebGIS is the province.Secondary data were taken from the shapefile of provinces in Indonesia. The contents of the provincial data field consisting of 34 provinces.

2.2. Analysis and Design

The analysis process focuses on provincial polygon data (secondary data) related to existing provincial data in primary data. This stage produces database design and use cases.The results of this design will be the basis for system development in WebGIS.

2.3. System Development

System development focuses on WebGIS based applications. The results of designing databases and use cases are the foundation in this stage. The database structure is built using MySQL. While the web-based application is built using HTML5, CSS, JavaScript, and PHP. The base map uses Google Maps.

Primary and secondary data are converted into CSV files. Then this file is imported into the database according to the structure in MySQL. WebGIS applications and databases are dynamically linked.

IOP Conf. Series: Materials Science and Engineering **990** (2020) 012003 doi:10.1088/1757-899X/990/1/012003

2.4. System Testing
System testing stages are based on system functionality. Testing on the WebGIS application using several browsers that are most widely used by users. System testing from the application side uses black-box functionality testing in accordance with ISO 9126-2 with 3 (three) matrices to be tested, namely Functional Adequacy (FA), Functional Implementation Coverage (FIC), and Functional Implementation Completeness (FICM).

2.5. Evaluation and Conclusion
Based on the data and distribution maps of the webinar participants on WebGIS, evaluations, and conclusions can be drawn about communication interests and learning motivation of participants. The parameters are the province and the professional region.

3. Result and Disscussion
The results and discussion are described in the technology schema, use cases, database structure, distribution maps, system testing, and the percentage of results from the dataset.

3.1. Technology Scheme
Technological schemes describe the relationship between data and systems. Data is accessed via the WebGIS application based on requests from the user to the server via the internet network. The data sources come from primary and secondary data in CSV format which have been imported into the MySQL database server. The technology schematic process flow is shown in Figure 2.

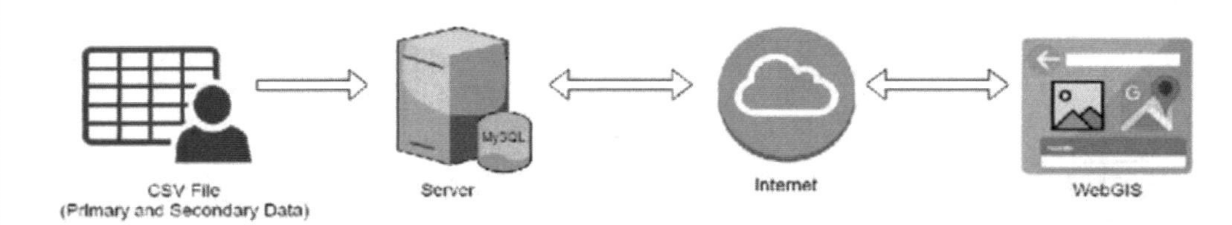

Figure 2. Technology scheme.

3.2. Use Cases
Use cases describe the relationship between the system and the user. Users will get information on participants who take part in the webinar based on their province. The system on the WebGIS application will display provincial info along with the number of participants (Figure 3).

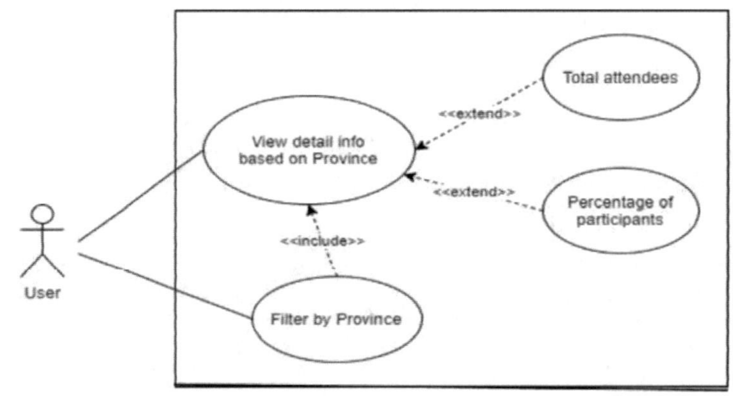

Figure 3. Use cases.

IOP Conf. Series: Materials Science and Engineering **990** (2020) 012003 doi:10.1088/1757-899X/990/1/012003

3.3. Database Structure

Based on the fields and contents of the webinar registration data, a database structure is created as the basis for database design. There are two main components that are used as tables in the database, namely participants and provinces. These two tables are interconnected by linking id_province as a Foreign Key (FK) in the participant table. This relation uses one-to-many.

The participant table itself has several fields as data entry fields including id_participants as Primary Key (PK), id_province as Foreign Key (FK), email, full name, age, profession, institution, and address. While the province table has several fields, namely id_province as Primary Key (PK), province_name, polygon, and total_attendees. The database structure is illustrated in Figure 4.

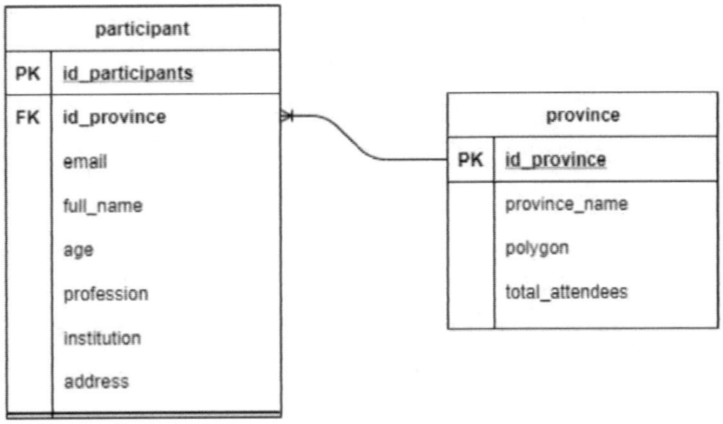

Figure 4. Database structure.

3.4. Distribution Map

The information distribution map or WebGIS of web-based webinar participants is accessed via a browser. In accordance with the technology scheme and use cases, users can access the number of attendees through a map of Indonesia based on their province. In addition, users can also see the percentage amount when clicking on polygons or provincial areas which appear as in Figure 5.

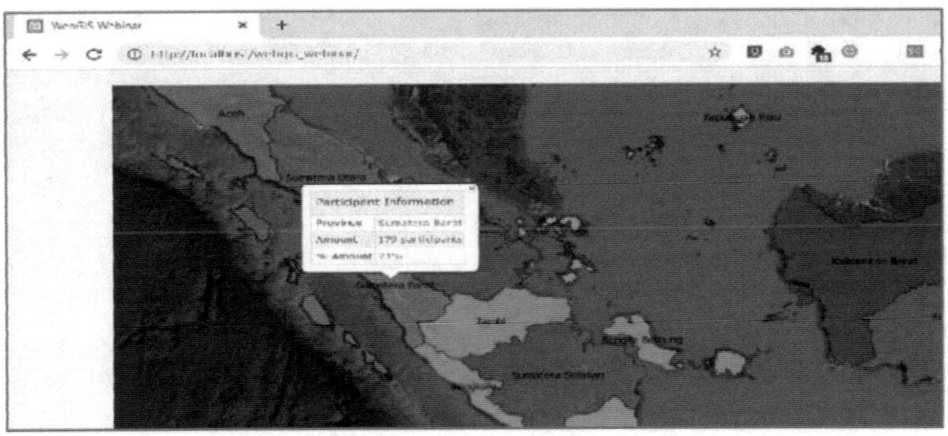

Figure 5. Distribution map.

3.5. System Testing

Based on the WebGIS application, the system testing process is carried out based on the four features in it. These four features are the main test cases in system testing. The test case values in the ISO 9126-2 test standard are in the range of 0 – 1. With a value of 0 (zero) it is said to be not good and a value close to or equal to 1 (one) is said to be good ($0 \leq X \leq 1$).Based on Table 1, it can be seen that all tests have a value of 1, meaning that the features in the WebGIS application can be run well and information on the number and percentage of participants can be seen through the provincial area on the distribution map.

Table 1. Functionality Testing Results.

No.	**Feature**	FA	*FIC*	FICM
1	**Displays the base map**	1	1	1
2	**Displays a map of the province**	1	1	1
3	**Filter by province**	1	1	1
4	**Displays info on the number of participants per province**	1	1	1

3.6. Percentage of Webinar Participants

The number of participants who registered for this webinar was 859 people. Participants are spread across several provinces throughout Indonesia. The total percentage of these three professions is 87.2%. The rest are school principals, employees, civil servants, and others.

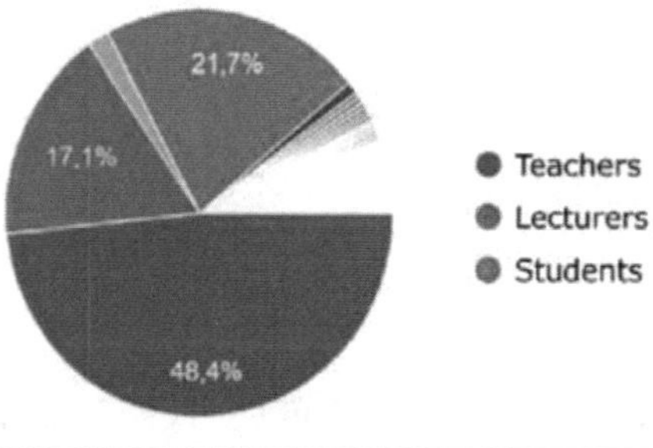

Figure 6. Percentage of Webinar Participants by Profession.

Of the 70 professions selected by the participants, there are three main professions that are related to the theme of this webinar. These professions are teachers, lecturers, and students. The highest percentage was obtained by teachers at 48.4%, students 21.7%, and lecturers 17.1% (Figure 6). Meanwhile, if based on the provincial area, the highest percentage was obtained by West Sumatra at 21%, East Java 16.4%, and West Java 10% (Figure 7).

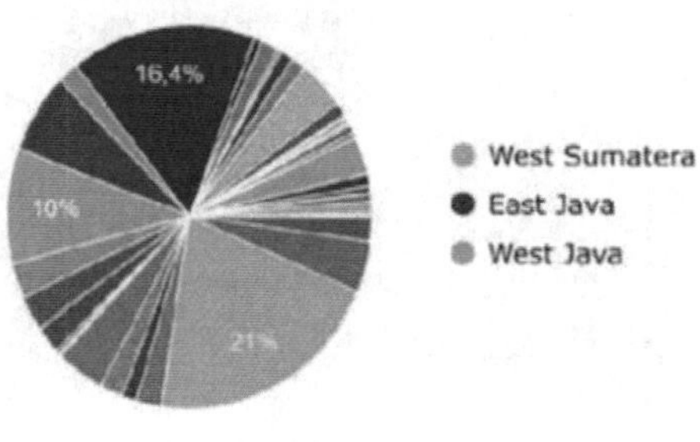

Figure 7. Percentage of Webinar Participants from Provincial Areas.

IOP Conf. Series: Materials Science and Engineering **990** (2020) 012003 doi:10.1088/1757-899X/990/1/012003

4. Conclusion

Based on the results of application implementation and data analysis, the participation of webinar participants on communication and learning motivation during the Covid-19 pandemic can be mapped via WebGIS. Of the 859 participants spread across Indonesia, the highest percentage was West Sumatra with 21%, East Java 16.4%, and West Java 10%. Area, amount, and percentage information can be accessed visually through the WebGIS application.

This participant distribution map information can be an illustration for universities, institutions, and stakeholders in the education sector in formulating policy and planning programs to improve the quality of teaching and learning through communication and learning motivation, both for teachers and students.

References

[1] Á. R. López, J. E. Souto, and M. L. A. Noblejas, "Improving teaching capacity to increase student achievement: The key role of communication competences in Higher Education," *Stud. Educ. Eval.*, vol. 60, no. October, pp. 205–213, 2019.

[2] Y. Daniarti, R. Taufiq, and B. Sunaryo, "The Implementation of Teaching Reading Through Genre Based Approach for University Students," *J. Phys. Conf. Ser.*, vol. 1477, no. 4, pp. 0–6, 2020.

[3] S. A. Mokhtar, Z. Zulfadli, S. M. S. Anuar, and S. Puteh, "Information system model for the measurement of learning outcome attainment," *2014 4th Int. Conf. Eng. Technol. Technopreneuship, ICE2T 2014*, vol. 2014-August, pp. 37–40, 2015.

[4] R. Djalante *et al.*, "Review and analysis of current responses to COVID-19 in Indonesia: Period of January to March 2020," *Prog. Disaster Sci.*, vol. 6, p. 100091, 2020.

[5] A. Patricia, "College Students' Use and Acceptance of Emergency Online Learning Due to COVID-19," *Int. J. Educ. Res. Open*, p. 100011, 2020.

[6] K. Murdaugh, J. Bainac Hausknecht, and C. T. Herbst, "In-Person or Virtual? -- Assessing the Impact of COVID-19 on the Teaching Habits of Voice Pedagogues," *J. Voice*, vol. in press, 2020.

[7] V. N. Anwar, "THE INFLUENCE OF PSYCHOLOGY OF MATHEMATICS EDUCATION ON STUDENT ' S CHARACTER IN ELEMENTARY SCHOOL," in *Proceeding the 2nd SEA-DR*, 2014, no. 978, pp. 374–381.

[8] D. L. Mishra, D. T. Gupta, and D. A. Shree, "Online Teaching-Learning in Higher Education during Lockdown Period of COVID-19 Pandemic," *Int. J. Educ. Res. Open*, p. 100012, 2020.

[9] P. D. MacIntyre, T. Gregersen, and S. Mercer, "Language teachers' coping strategies during the Covid-19 conversion to online teaching: Correlations with stress, wellbeing and negative emotions," *System*, vol. 94, p. 102352, 2020.

[10] Budi Sunaryo *et al.*, "Mapping Mining Potential Using WebGIS," *SciTech Framew.*, vol. 1, no. 1, pp. 41–46, 2019.

[11] J. F. Rusdi *et al.*, "Dataset smartphone usage of international tourist behavior," *Data Br.*, vol. 27, 2019.

IOP Conf. Series: Materials Science and Engineering **990** (2020) 012004 doi:10.1088/1757-899X/990/1/012004

Decay Performance Modelingof Historic Timber Roof Structures of Amantubillah Palace in Mempawah, Indonesia

HEPutra[1], HPrabowo[2,*],DRIndriana[1]

[1]Department of Architecture, PoliteknikNegeri Pontianak, Jl. Jenderal Ahmad Yani, Kota Pontianak 78124, Kalimantan Barat, Indonesia

[2]Laboratory for Construction, Innovative Structures, and Building Physics,PoliteknikNegeri Pontianak, Jl. Jenderal Ahmad Yani, Kota Pontianak 78124, Kalimantan Barat, Indonesia

*prabowoherry@yahoo.com

Abstract. Durability problem is a usual shortcoming of utilizing timber as a construction material. After exposed to the weather in decades, the historic timber building often experienced a severe deterioration process. Amantubillah Palace, as Mempawah Sultanate's inheritance, also suffers from this process.The repair and replacement action are needed to preserve this legacy from the past.Thus, it is essential to know the remaining service life of the building. The calculations of decay depth were carried out by utilizing the Australian CSIRO model incorporating the parameters which are related to the Indonesian tropical climate. From the calculations, it is obtained that the decaying process will reach the centre of the cross-section in 83 years. It is recommended that repair or replacement action should be taken immediately.

1. Introduction

Amantubillah Palace is located at Mempawah Regency, West Kalimantan Province, Indonesia. It was first built during the reign of GustiJamiril, who has the title PanembahanAdiWijayaKesuma (1761-1787). He was the 3rd Sultan of Mempawah Sultanate. Amantubillah Palace caught fire in 1880. It was rebuilt and in function again on November 2nd, 1922, when Gusti Muhammad TaufikAccamaddin (1902-1943) reigned Mempawah.

The Palace is a timber building made of local material. In the past, timber is abundant and easily found in the tropical forest of Kalimantan. Many timber species haveexcellent durability properties as a construction material. Due to this reason, most of the historic building in West Kalimantan is timber building.

The common drawback of using timber as a construction material is its durability issues. The preservation actions such as repair and replacement of the main building elements should be doneat the right time. Thus, it is vital to investigate the occurring deterioration process and predict the building'sremaining service life. Numerous research has been done in simulating the service life of the timber structures.The combination of durability with strength parameter had been offered by Van de Kuilen [1], building physicshygrothermal model had been studied by Viitanen et al.[2], and damage functions coupled with structural and hygrothermal model had been proposed by Nofal and

[1] prabowoherry@yahoo.com .

Content from this work may be used under the terms of the Creative Commons Attribution 3.0 licence. Any further distribution of this work must maintain attribution to the author(s) and the title of the work, journal citation and DOI.
Published under licence by IOP Publishing Ltd

Kumaran[3]. Corrosion of metal fasteners embedded in the solid wood model had been simulated by Zelinka et.al [4], and the decaying process model combined with hygrothermal analysis was studied by Saito at.al [5]. Massive and extensive field and laboratory research in Australia had been shared by Wang at.al[6] and MacKenzie at.al[7]. A dose-response performance model had been shown by Bornemann at.al [8], and wood durability and preservation in European Standardization was presented by Kutnik at.al[9]. A dose-response performance using the factorization approach was introduced by Meyer-Veltrup at.al[10], and a decay model incorporating Indonesian tropical climate parameters was presented by Prabowo&Hilmy[11].

The durability model introduced by MacKenzie at.al [7]will be addressed due to limited timber durability research in Indonesia. The regional nearness is one of the reasons to utilize the model. The parameters included in the model are Timber durability class and species of local timber, temperature, relative humidity, and degradation mechanism.

2. Methodology

The decay performance of roof structures will be evaluated by taking the CSIRO (Commonwealth Scientific and Industrial Research Organisation). It is based on the Australian climate condition[7]. The decay above ground condition will be addressed for the model. The roof structures' components will be measured for their thermo-hygric properties. As a comparison, local relative humidity and temperature data from the nearest weather station will be collected and plotted. The dimension of the building and roof structures taken from field measurements will be given.

2.1. The model of decay depth

The model of decay depth constitutes two different components. The first component describes the decay depth that occurred before and at the time when deterioration initiates. After the initiation of deterioration, the second component of the decay depth model will be used.

The decay depth model includes two controlling parameters: a decay lag (tlag) and a decay rate (r). A decay lag is measured in a unit of years where a decay rate is quantified in a unit of mm/year. All related parameters to decay depth model are expressed in equation (1) to equation (5), where d(t) = decay depth after t years of installation, d(0) = decay depth at the initiation time, k_{wood} = wood parameter, $k_{climate}$ = climate parameter. The model of decay depth is demonstratedin figure 1.

$$d(t)=\begin{cases} ct^2 & \text{if} \quad t \le t_{d_0} \\ (t-t_{lag})r & \text{if} \quad t > t_{d_0} \end{cases} \tag{1}$$

$$t_{d_0} = t_{lag} + \frac{d_0}{r} \tag{2}$$

$$c = \frac{d_0}{t_{d_0}^{\,2}} \tag{3}$$

$$t_{lag} = 8.5r^{-0.85} \tag{4}$$

$$r = k_{wood}k_{climate} \tag{5}$$

IOP Conf. Series: Materials Science and Engineering **990** (2020) 012004 doi:10.1088/1757-899X/990/1/012004

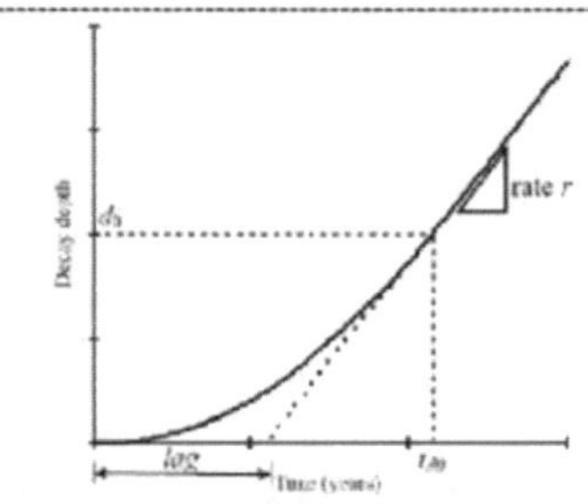

Figure 1. The model of decay depth.

2.2. Local relative humidity and temperature

The relative humidity and temperature datawere obtained from the nearest weather station. The data were recorded by Mempawah Climatological Station, Mempawah Regency, Indonesia. They are collected from the year 2004 until 2019 to simulate the weather condition in July[12]. The relative humidity (RH) and temperature (T) are illustrated in figures 2 and 3.

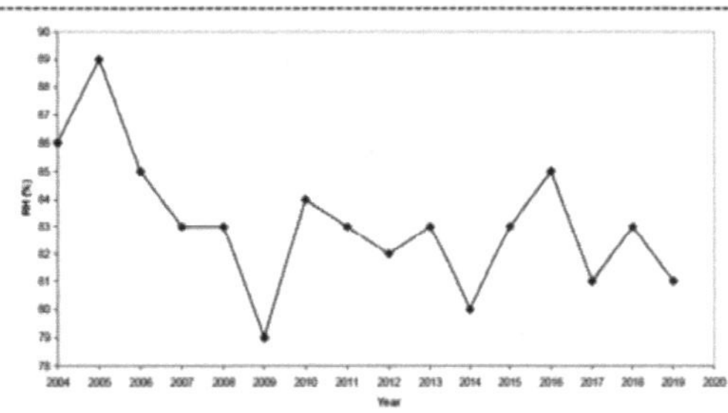

Figure 2. RH in july from 2004 to 2019.

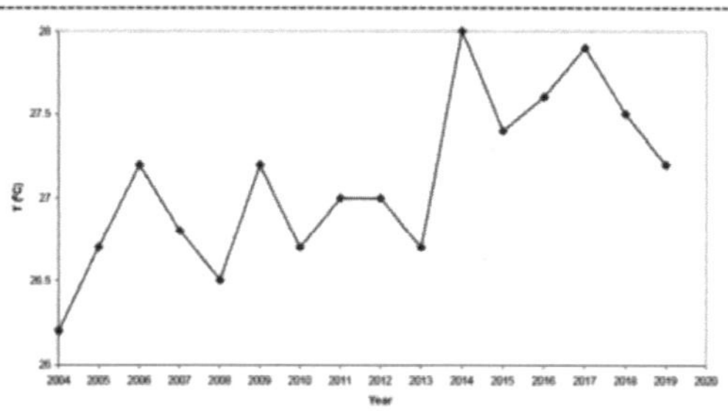

Figure 3. T in july from 2004 to 2019.

2.3. The roof structures

The roof structures are in the form of timber trusses. The trusses are made of two local timber species: Belian (ulin) (Botanical name: *Eusideroxylonzwageri*) and Bangkirai (Botanical name: *Shorealaevis*). These two species are well-known for their durability against the weather. They are classified as durability class 1 and durability class 2, respectively. The cross-section of the truss is rectangular with dimension 8 cm x 14 cm. The truss members are jointed with a wooden dowel. The roof truss can be seen in figure 4.

The main building width and length are approximately 36 m x 47 m. The main components of the building are made of Belian. The main building consists of the throne room, main rooms, and the porch. The plan and front view of the buildingcan be noticed in figure 5.

Figure 4. The roof truss.

IOP Conf. Series: Materials Science and Engineering **990** (2020) 012004 doi:10.1088/1757-899X/990/1/012004

Figure 5. The plan and front view of the building.

2.4. *The thermo-hygric evaluations*

The thermo-hygric evaluations of the roof structures' components were carried out by utilizing a humidity-temperature digital instrument (YK-2001 TM type, Lutron, Taiwan), as shown in figure 6. The measurements referred to ASTM F2420-05 [13]. It is conducted three times in one day: morning (8 am to 10 am), noon (11.30 am to 1.30 pm), and afternoon (4 pm to 6 pm). The measured parameters are graphed in figure 7 and figure 8.

Figure 6. The thermo-hygric evaluation.

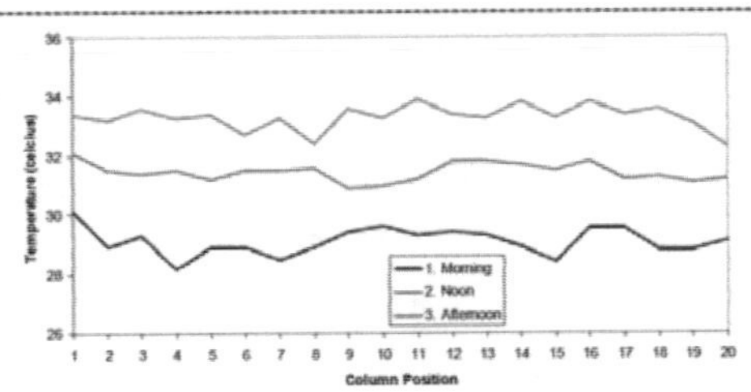

Figure 7. The evaluated temperature.

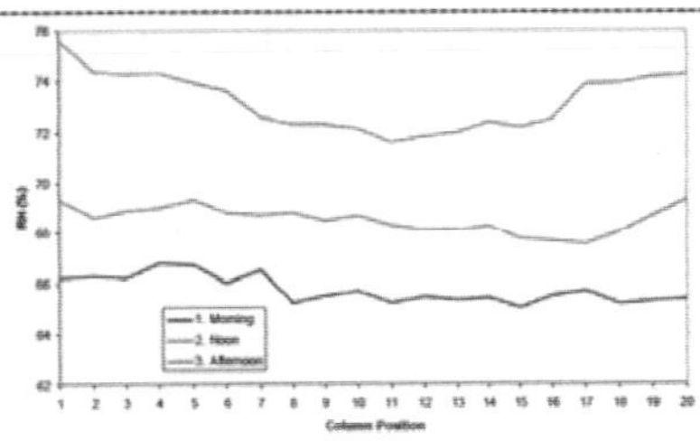

Figure 8. The evaluated relative humidity.

3. Results and Discussions

The decay depth calculations have been performed to simulate the service life of the roof structures.The assumed decay depth initiation is 5 mm under the surface of roof structures' cross-sections. The model input parameters are k_{wood} = 0.5 (ulin species), k_{wood} = 0.62 (bengkirai species), $k_{climate}$ = 0.4 (the least hazardous zone), k_p = 3.5 (ulin, painted timber), k_t = 1 (thickness 80 mm), k_w = 1,1 (width 80 mm), k_n = 1 (no connector), k_{g1} = 0.3 (non-contact), k_{g2} = 2.

The depth decay calculation during the building service life can be seen in figure 9. It illustrates the d(t) vs. t, especially for building service life, which is almost 100 years. The graph shows that a decay rate (r) = 0.46 mm/year for Belian and r = 0.57 mm/year for Bangkirai. The decay lag (tlag) is equal to

16.389 years for Belian and 13.648 years for Bangkirai. These values mean the initiation of deterioration in 16 years for Belian and 13 years for Bangkirai right after the building was built. The first component of the graph shows a parabolic curve referred to as equation dt = ct2. A relatively linear curve is simulated by the second component of the graph under the equation (t-tlag)r.

The initiation of tlagcan be well observedinfigure 10. It is graphed to show the initiation of tlag more clearly. The graph shows that after 40 years, the decay depth reaches 11 mm for Belian and 15 mm for Bangkirai. The decay depth will reach 40 mm in 100 years for Belian and 83 years for Bangkirai.

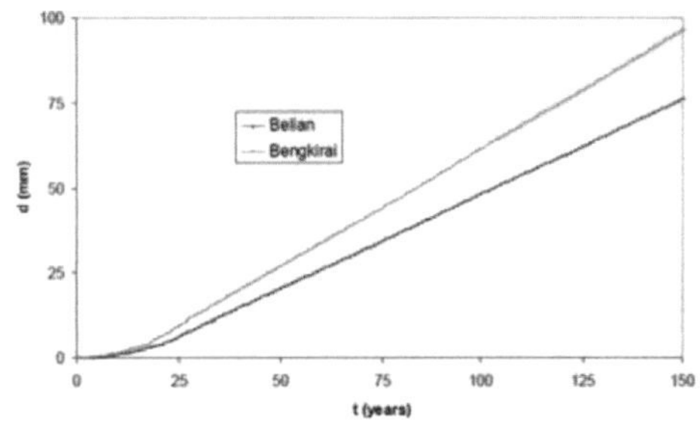

Figure 9. The decay depth during the building life.

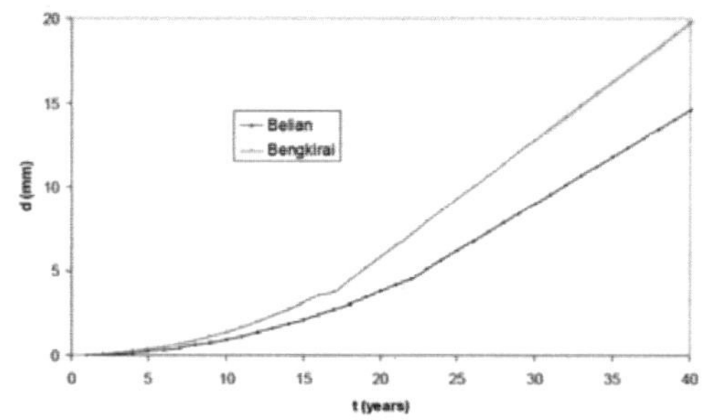

Figure 10. The initiation of tlag.

4. Conclusion

The demand forthe simple and practical formulation of service life prediction of a historic building is increasing. It is needed to determine the repair or replacement time of the deteriorated building components. The calculations of decay depth give that a decay rate (r) = 0.46 mm/year for Belian and r = 0.57 mm/year for Bangkirai. The decay lag (tlag) is equal to 16.389 years for Belian and 13.648 years for Bangkirai. The decaying process will reach the centre of the cross-section in 100 years for Belian and 83 years for Bangkirai. This condition will risk the stability and safety of the roof structures. It is advised that the repair action, such as adding area or replacement of the existing cross-section, should be implemented.

Acknowledgment

Authors wish to acknowledge financial support from DIPA Polnep fiscal year 2020. The assistance or encouragements from colleagues and extraordinary works by technical staff are also gratefully acknowledged.

References

[1] van de Kuilen JW 2007 *Materials and Structures* 40.

[2] Viitanen H, Toratti T, Makkonen L, Peuhkuri R, Ojanen T, Ruokolainen L 2016 *European Journal of Wood and Wood Products*68.

[3] Nofal M, Kumaran K 2011 *European Journal of Wood and Wood Products* 69.

[4] Zelinka SL, Derome D, Glass SV 2011 *Building and Environment*46.

[5] Saito H, Fukuda K, Sawachi T 2012*Building Simulation* 5.

[6] Wang CH, Leicester RH, Nguyen MN 2008*Manual 4 – Decay above- ground. Victoria: Forest and Wood Products Australia Limited (CSIRO Sustainable Ecosystems).*

[7] MacKenzie C, Wang CH, Leicester RH, Foliente GC, Nguyen MN 2013 *Timber service life design - Design guide for durability (Revised Version)*. Victoria: Forest and Wood Products Australia Limited.

[8] Bornemann T, Brischke C, Alfredsen G 2014 *Wood Material Science and Engineering*9.

[9] Kutnik M, Suttie E, Brischke C 2014 *Wood Material Science and Engineering* 9.

[10] Meyer-Veltrup L, Brischke C, Niklewski J, Hansson EF 2018 W*ood Material Science & Engineering*13.

[11] Prabowo H, Hilmy M2019 *IOPConf. Series: Earth and Environmental Science*; IOP Publishing.

[12] BPS. Provinsi Kalimantan Barat dalam Angka 2020. Pontianak: Badan Pusat Statistik.

[13] ASTM. ASTM F2170 - 16b 2016 *Standard Test Method for Determining Relative Humidity Using in situ Probes*. West Conshohocken, PA

IOP Conf. Series: Materials Science and Engineering **990** (2020) 012005 doi:10.1088/1757-899X/990/1/012005

Load Shedding Simulation Using A Frequency Relay In Lampung Electrical System

I Darmana[1], A R Salvayer[2], Erliwati[3]
[1]Electrical Engineering,Universitas Bung Hatta Padang,Indonesia.
[2]Graduated from Electrical Engineering, Universitas Andalas Padang, Indonesia.
[3]Electromedical Engineering, Politeknik Kesehatan Siteba Padang, Indonesia.

E-mail: ija.darmana@bunghatta.ac.id

Abstract,Stability is one of the requirements of an electrical power system which can be considered good/ideal, One disturbance that can affect stability is the outage of the system generator. To overcome this problem, load shedding is performed, The load is released in a simulation aimed at bringing the frequency back to the authorized value after the disturbance has occurred. The power plant that experienced a trip in this study was the largest Lampung unit of the generator Tarahan 4.When the generator Tarahan 4 has been tripped, the frequency system is falls and stabilize at 48.8 Hz.Thereforeits necessary to release the load so that the system frequency returns to the authorized value, which is from 49,8 Hz to 50.2 Hz. Based on the results of simulation after setting the frequency relay, by releasing a load of 152.2 MW, the frequency was stable at 50.074 Hz . So it can be concluded that the frequency relay setting has been able to return the system frequency to allowable value.

1.Introduction

If a generator trips, then the power that was originally generated turns into a lack of production power, this will cause the frequency to drop. If these conditions are not resolved immediately, this may cause a system failure due to the operation of the under frequency relay, and generators operating at frequencies which do not comply with their working frequency limits, will cause excessive vibration in the turbine and can shorten the life of the generator [1]. It is therefore necessary to conduct a study by applying a simulation, this study entitled: "*Load Shedding Simulation Using a Frequency Relay in Lampung Electrical System*".

The load shedding must be carried out as optimal as possible so that the released load is minimum[2].Load shedding has several methods, including: load shedding using the Under Frequency Relay (UFR) and Under Voltage Load Shedding (UVLS); Unloading applies the semi-adaptive method, using the Rate of Change of Frequency (RoCoF) relay [3-4]; The load shedding applies the adaptive method.In this method lies in adding the frequency response or called the system frequency response (SFR), SFR contains information such as the governor's response,

---

[1] ija.darmana@bunghatta.ac.id

Content from this work may be used under the terms of the Creative Commons Attribution 3.0 licence. Any further distribution of this work must maintain attribution to the author(s) and the title of the work, journal citation and DOI.
Published under licence by IOP Publishing Ltd

1.1. Electric system of Lampung province
The electricity system of Lampung province is a system interconnected with the electricity system of South Sumatra, so that if Lampung lacks generation, it will benefit from a transfer of electricity from South Sumatra. However, the power transfer has a maximum transfer limit so that the system is still able to maintain its voltage stability (PV curve). In addition, the limited power transfer is also influenced by the MVA nominal value of the interconnected generator. On the other hand, Lampung is very dependent on energy transfer from South Sumatra because the Lampung power plants have not been sufficient to meet the load needs of the region.

At the peak of load on November 5, 2019, the transfers from South Sumatra to Lampung are very high, reaching 336 MW (the highest power transfer on the island of Sumatra). This means that production at Lampung was not sufficient to meet the needs of the load in their own area, so it requires a significant transfer of energy from South Sumatra. Under these conditions, the Lampung electrical system is subject to stability caused by disturbances. The disturbance can be in the form of a generator trip which results in a decrease in system frequency [4]. If the impact is not overcome, it will cause a total breakdown. Therefore, we need a defense system that can prevent total breakdown in the Lampung electrical system. In this study, the authors will apply the charge release scheme using the semi-adaptive method.

2. Research objectives
The objectives of this research are:
- Know the state of the system when the trip is one of the largest power plants in the Lampung subsystem
- Determine the parameters of the RoCoF relay and the amount of charge released
- Analysis of the state of the system after the RoCoF relay has operated

3. Research methods
In this research, the author uses Digsilent Power Factory to analyze the problems observed, Steps to follow before obtaining the RoCoF relay [3] parameters: 1) Data collection, 2) Single line diagram modelling, 3) Power flow simulation, 4) Short circuit simulation, and 5) Generator interference simulation trigger.

4. The Data
Here is the data necessary to build a single-line diagram of the province of Lampung

4.1.Data Generator

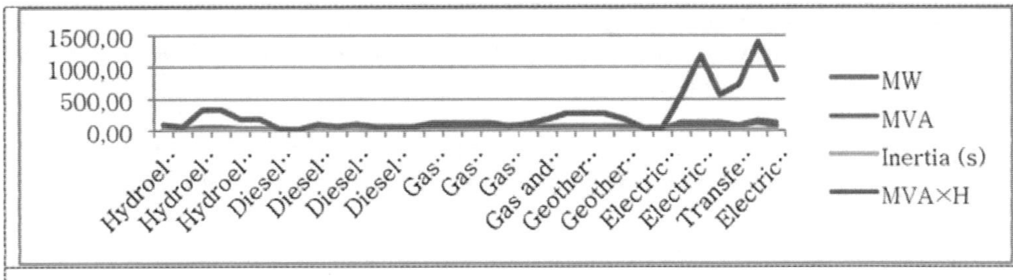

Figure 1. Data Generator1 to 33

4.2. Load data for Substation

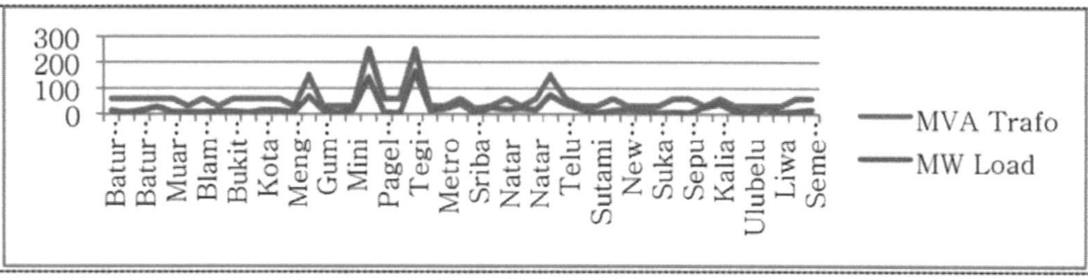

Figure 2. Load Data for Substation 1 to 47

4.3 Single Line Diagram of Lampung Electric System

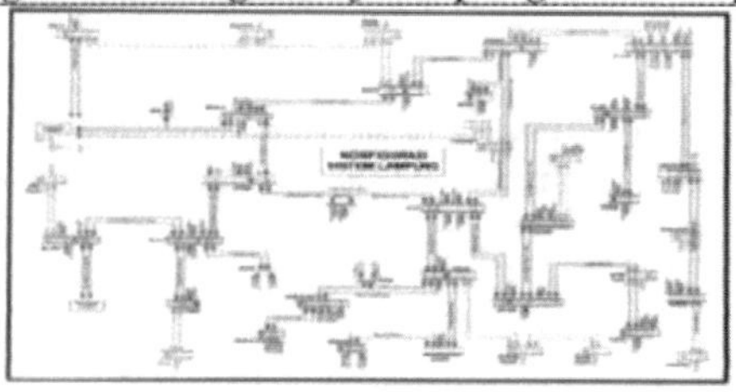

Figure 3. Single Line Diagram of Lampung SubSystem

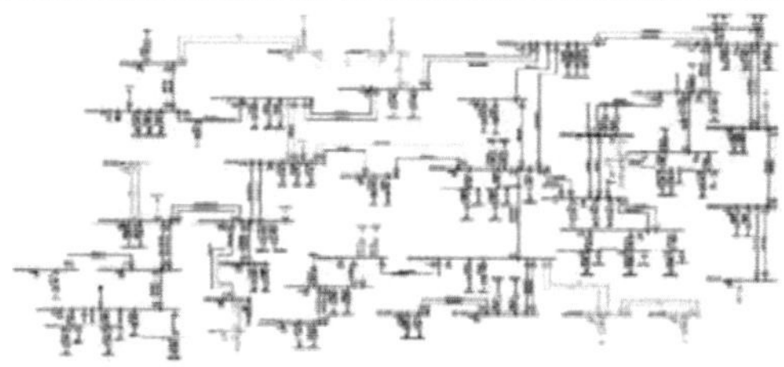

Figure 4. Single Line Diagram Modeling using Digsilent Power Factory 15.1

5. Simulation

5.1. Power flow simulation

The power flow simulation aims at verifying the data entered in the system components (such as the generators and the loads) according to the real data of the Lampung subsystem [5-6].

5.2. Simulation of a short circuit interruption

This simulation aims to check if the circuit is correctly connected [5-6], A three-phase short-circuit disturbance is given on the transmission line 1 which connects Tegineneng Substation and Pagelaran Substation.

5.3. Simulation of trip generator interference

This simulation aims to determine the condition of the system frequency after the generator experiences trip [5-6]. The power plant that experienced a trip in this study was the largest generator Tarahan 4. With a trip time occurring in the 1st second after the simulation is run. The following is the system frequency Fig.5.

When the generator Tarahan 4 has been tripped, the frequency system is falls and stabilize at 48.8 Hz.So it is necessary to release the load so that the system frequency returns to the authorized value, which is from 49.8 Hz to 50.2 Hz.

6. RoCoF relay parameters

To obtain the inertia of the system using the formula:

$$H_{sys} = \sum_{i=1}^{n} \frac{H_{generator} x\, MVA_{generator}}{MVA_{generator}} \quad (1)$$

IOP Conf. Series: Materials Science and Engineering **990** (2020) 012005 doi:10.1088/1757-899X/990/1/012005

In this case study, the trigger generator is the largest generator, namely Steam Power Plant of Tarahan 4. To obtain the correct system inertia in determining df/dt, the total system inertia must be reduced by inertia of the disturbed unit [3], so that$\boldsymbol{H_{sys}}$= **5.2844.**Determine the apparent power (MVA) and the value of the active power system (MW) shortly after the disruption of Tarahan steam power plant 4 [7]. Based on Figure 6, the following values are obtained: MW generated by generator = 732.623 MW, MVA generated by generator = 739.867 MVA, MW generated by external network (transfer from South Sumatra) = 221.552 MW, MVA generated by external network (transfer from South Sumatra) = 236.96 MVA, MW load = 941.803 MW.

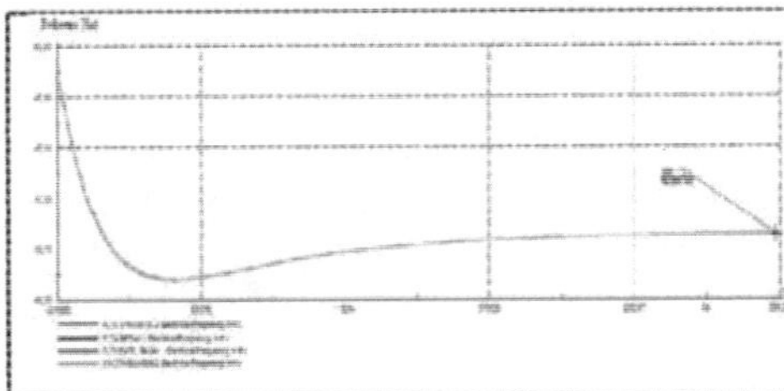

Figure 5. Frequency of the system after trip on Tarahan Steam Power Station 4

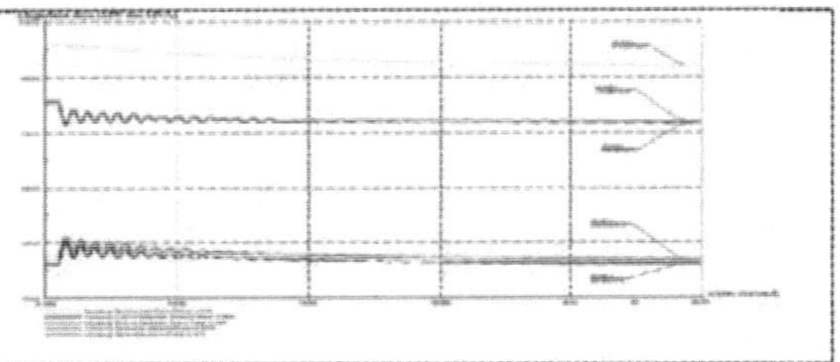

Figure 6. Chart of MW and MVA of Lampung system after the trip of Tarahan Steam PP 4

After knowing the MW and MVA system, the df/dt calculation is performed = - 0.06 Hz/s.For the RoCoF relay time delay taken at 0,05 seconds, this value is taken from a range of values from 50 to 500mS. This value is within the RoCoF relay tolerance limit to prevent trip errors [3].
The decrease in system frequency is obtained from the ratio between the power supplied by the unit that is interrupted before experiencing disruption to the remaining installed power in the system, df/dt = -0.346 Hz/s.The number of MW of load that must be released by the RoCoF relay [3] can use the formula:

$$\left|\frac{df}{dt}\right| = \frac{f_s}{2H}\left\{\frac{P_{BU}-(P_{SU}-P_{BR})-P_B}{P_{GOT}-P_{SOT}}\right\} \quad \textbf{(2)}$$

Where:

df/dt :Expected changes in frequency after the load is released

Based on simulation, it is known that the RoCoF relay works when the frequency reaches 49.962 Hz and is at 1.405 seconds.So that the PB values are obtained = 941.08 MW.To determine how many MW must be released, obtained = 159.3 MW.So, by releasing a load of 159.3 MW, it is expected that the system frequency will return to the value of 50 Hz.After releasing the load, the frequency had increased to 50.685 Hz at 31.266 s seconds. After that the frequency is stable at 50.074 Hz.

7. Analysis of Simulation Results

The increase in frequency which reaches a value of 50.685 Hz is due to the fact that the system responds to a significant reduction in the MW load due to the operation of the set RoCoF relay [3]. So at that time, the MW load is smaller than the MW generated. But the increase occurred in a short time, so it is still within the frequency tolerance range. After a period of frequency increase, the system frequency is constant at 50.074 Hz,it show at figure 7. Indeed, the conditions for generating MW are proportional to the load MW. The value of 50.074 Hz is a value that is always within the safety frequency limit, where the maximum frequency is 50.2 Hz to be able to operate continuously.

IOP Conf. Series: Materials Science and Engineering **990** (2020) 012005 doi:10.1088/1757-899X/990/1/012005

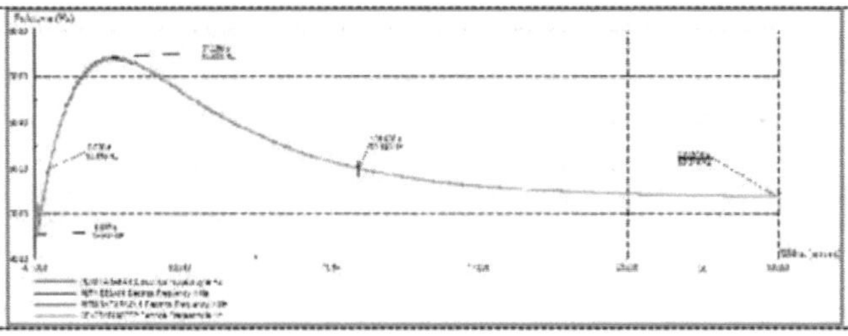

Figure 7. Frequency System After Load Shedding is Occur

8. Conclusion

Based on the analysis of the data and observations during the practical work, the following conclusions are obtained:

1. Lampung's electrical system is subject to stability which can cause power outages. In fact, the frequency decreases when the largest generator trips. On the other hand, adding power plants requires high costs, so defense plans are needed to maintain the reliability of the system.
2. The RoCoF relay pays attention to the power imbalance when the generator is trip.
3. To determine the number of MW of load released, the increase in frequency df/dt must be equal to the decrease in frequency df / dt,but it has a different polarity. The frequency is expected to increase and return to its nominal frequency (50 Hz).
4. The use of RoCoF relays can be used as a defense system for the Lampung system because according to the results of the study.This relay is capable of releasing the load and reaching the authorized frequency.

Acknowledgments

On this occasion, the author wishes to express his gratitude and thanks to all those who greatly helped to carry out this research, in particular to the president and operational staff of the UIP3BS system, Pekanbaru.

References

[1] Zhang, Guorui, 2009, *EPRI Power Systems Dynamics Tutorial*, Electric Power Research Institute, California.

[2] Bakar, Nur Najinah, et al, 2016, *Microgrid and Load Shedding Scheme During Islanded Mode : a Review*, Elsevier.

[3] Turaj Amraee, Alireza Soroudi at al, 2017, *Probabilistic Under Frequency Load Shedding Considering RoCoF Relays of Distributed Generators*, IEEE, DOI 10,1109/TPWRS,2017,2787861.

[4] Alexander Berdin and Pavel Kovalenko, 20018,Defining The Power System Load Frequency Static Response Slope Based On Transient Synchrophasor Data, MATEC Web of Conferences 208, 04002 (2018) https://doi,org/10,1051/matecconf/201820804002.

[5] Putu Agus Aditya Pramana, Buyung Sofiarto Munir, 2016, *Modeling and Simulation of Electrical Resonance in EHV Transmission Line Case Study in West Java Region 500 kV System*, MATEC Web of Conferences 55, 05005 (2016), DOI: 10,1051/matecconf/20165505005, ACPEE.

[6] Nikolay Yu, Ruban, AlmazO Sulaymanovat al, 2017, *Simulation Of A High-Frequency Link Of Phase Comparison Protection Of Transmission Lines For Optimization Its Settings*, MATEC Web of Conferences 91, 01050 (2017), DOI: 10 ,1051/matecconf/20179101050, Smart Grids 2016.

[7] Marsudi, Djiteng, 2006,*Operasi Sistem Tenaga Listrik,* second edition, Graha Ilmu Publisher.

IOP Conf. Series: Materials Science and Engineering **990** (2020) 012006 doi:10.1088/1757-899X/990/1/012006

Effect of Thickness and Type of Magnet against EMF Back PMSG 12S8P with FEM

Liliana[1], Z Aini[1], A Wenda[1], T D Putri[1]
[1]Electrical Engineering Departement, UIN Suska Riau Pekanbaru, Indonesia

E-mail: liliana@uin-suska.ac.id

Abstract. Synchronous Magnet Generator (PMSG) is one of the main components of wind power generation. This generator has excellent low speed applied especially for less strong winds. Electromotive Force (EMF) output is expected to be enhanced for higher PMSG efficiency. This research aims to analyze the influence of thick and magnetic types on EMF Back PMSG 12S8P. Thickness and type of magnets vary to get the highest EMF output. The thickness is variated at a value of 3 mm and 6 mm, while the magnetic material used Neodymium Iron Boron (NdFeB) and PM12 Br 1.2 Mur 1. Linkage flux value is generated with Finite ElemenMethode (FEM) using Magnet 7,5 Software for each variation. From analyze results and compared to other variations, the highest EMF value obtained from the variation in magnetic thickness 6 mm with material type PM12 Br 1.2 Myrrh 1 with EMF value reaching 21.13122 Volt.

1. Introduction

The Generator is one of the electrical machines that can turn the kinetic energy into electrical energy. We need a reliable, low speed generator with good efficiency, especially in the not-so-tight winds. The Alternator Permanent Magnet meets these qualifications. So far the generator is the most popular choice in designing a successful small windmill e.g. generator I-2 permanent magnet 18 slots 16 Pole (coggingless) with output generated 500 W with wind speed limitation of 12 m/s [1,3].

The magnitude of voltage greatly affects the power output in generators. The magnitude of the voltage depends on the speed of the rotation, the amount of wire in the coil that cuts the flux, and the magnitude of the magnetic flux raised by the magnetic field and the generator's gear [2]. The resulting flux is limited in a state of saturation (maximum condition) so as to operate at a low speed, the generator should work at a lower voltage. This can be done by changing the coil connection from parallel to a series or Delta to a star or wearing a more number of rounds of percoils [4].

Flux is produced by magnetic field from permanent magnet used. The permanent selection of magnets with the best quality is preferred to produce larger flux. The properties of permanent material magnets will directly affect the performance of the generators. Knowledge of the properties of good material is indispensable to be able to choose the right material. In recent year magnetic materials such as aluminium nickel and Cobalt (Alnico) alloys, StrosiumFerit or Barium Ferrite (Ferit), Samarium Cobalt (SmCo) are the first generation of Rare-Earth and Neodymium Iron-Boron Magnets (NdFeB) being the second generation. NdFeB is the most common Rare-Earth magnet used on current generators [4,5]. This type of magnet is known to be the best-in-class magnet. Other magnet also very

[1] liliana@UIN-suska.ac.id.

Content from this work may be used under the terms of the Creative Commons Attribution 3.0 licence. Any further distribution of this work must maintain attribution to the author(s) and the title of the work, journal citation and DOI.
Published under licence by IOP Publishing Ltd

good is the type of magnet PM12; Br 1.2 Mur 1. These two permanent magnets have the highest permanent magnet characteristics, have a maximum energy product value and also have a high cohesive force [7]. The thickness of the magnet itself will also affect the flux output, the larger the magnetic dimension then the flux spread will also be wider [6].

The study aims to analyse the influence of thickness and magnetic material against the EMF back output and K_E of PMSG 12S8P. The EMF back is caused by an ever-changing electric field cut on the conductor on the dynamo or generator. The terrain that continues to change on the dynamo conductor is caused or produced by the use of EMF. The EMF back value can be derived directly from the linkage flux that is quoted as Dψ using Faraday's law that is represented in the equation below [4].

$$E(\xi) = \frac{d\psi}{dt} = \omega_m = \frac{d\psi(\xi)}{d\xi} \quad (1)$$

where, $E(\xi)$ is EMF back (V), $d\psi$ is flux change, and dt is time span (s). The output of the EMF back relies on the strength of the magnetic flux line, the number of coil on the conductor, a magnetic flux intersecting angle with the conductor, and the velocity of the conductor in cutting magnetic flux lines. EMF back or K_E constants represent the construction of the machine. If all coils that are in one phase are connected and have the total peak value of EMF back then the K_E is defined in the following equation [4]

$$K_E = \frac{e_{ph}}{\omega_m} \quad (2)$$

where K_E is Konstanta EMF back, e_{ph} is average EMF (V), and ω_m is angular speed (rad/s). The simulation was made into four model variations namely models with a thick 3mm and 6mm magnet that uses the Neodymium Iron Boron material (NdFeB) and a model with a thickness of 3mm and 6mm magnets using material PM12: Br 1.2 mur 1.0. From this simulation result will be analysed combination which has the best EMF back output result on PMSG 12S8P.

2. Methodology

2.1 Research Prosedure

The procedure to get the best EMF Back value of PMSG 12S8P is presented complete as follows:

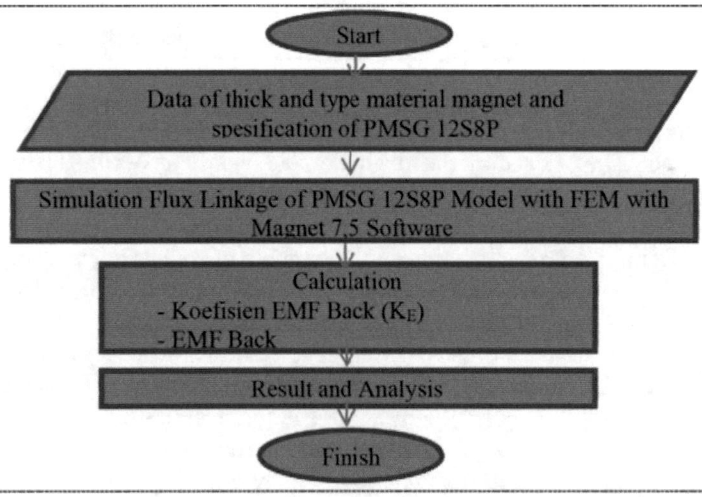

Figure 1. Research Procedure

IOP Conf. Series: Materials Science and Engineering **990** (2020) 012006 doi:10.1088/1757-899X/990/1/012006

2.2. Description Research Procedure

2.2.1. Research Data

Magnet permanent used with two thickness variations is 3 mm and 6 mm while the permanent magnet material to be used in this simulation is Neodymium Iron Boron (NdFeB) and PM12 Br 1.2 Mur 1

Table 1. Thick and type of permanent magnet variation

Variation	Thicknees (mm)	Material
1	3 mm	odymium Iron Boron(NdFeB)
2	6 mm	odymium Iron Boron(NdFeB)
3	3 mm	PM12 Br 1.2 mur 1.0
4	6 mm	PM12 Br 1.2 mur 1.0

Specification model PMSG 12S8P was used in the studies given below

Table 2. Specification PMSG 12S8P

Spesification	Description
Number of slots	12
Dimensions	180 x 180 x 40
Number of Poles	8
Number of Coil	10
Stator Material (iron Core)	Carpenter: Silicon steel
Coil Material	Copper: 5.77e7 Siemens/meter
Rotor Material	Carpenter: Silicon steel
Airbox Material	Air
Airgap Material	Air

2.2.2. Simulation Flux Linkage with FEM using Magnet 7,5 Software

The simulation steps of PMSG 12S8P to get linkage flux values are as follows : (1) Setting the display tools FEM on Magnet Software, (2) Create geometry design PMSG 12S8P, (3) Variation of thickness and type of magnet, and (4) Testing so as to get Flux Linkage

3. Results and Discussions

3.1. Flux Linkage

From the flux linkage data graph, it is known that the modelling of this generator is a 3-phase generator consisting of a phase U, phase V and phase W. This graph is a rotor rotation graph against the output flux value generated per 360 ° or by one round of the rotor, and it can be seen that the magnitude of flux with U, V, W coil at each angle change will result in flux forming sinusoidal waves caused by the change of north and south pole magnet through the coil during rotation rotor. The data to be presented is a flux output data against a 3° rotor rotation of up to a full rotation. For flux output of each variation model will be described as follows.

From the chart (d) of the material variation model of PMSG PM12: Br 1.2 Mur 1.0 with a magnetic thickness of 6 mm above can be seen that the flux is produced when the rotor rotates 360 ° empty touching the 0.008 line and this model results in the highest flux output among the three other models (a,b, and c). This Model proves that magnetic permanent material also greatly affects the flux generated in each rotor rotation.

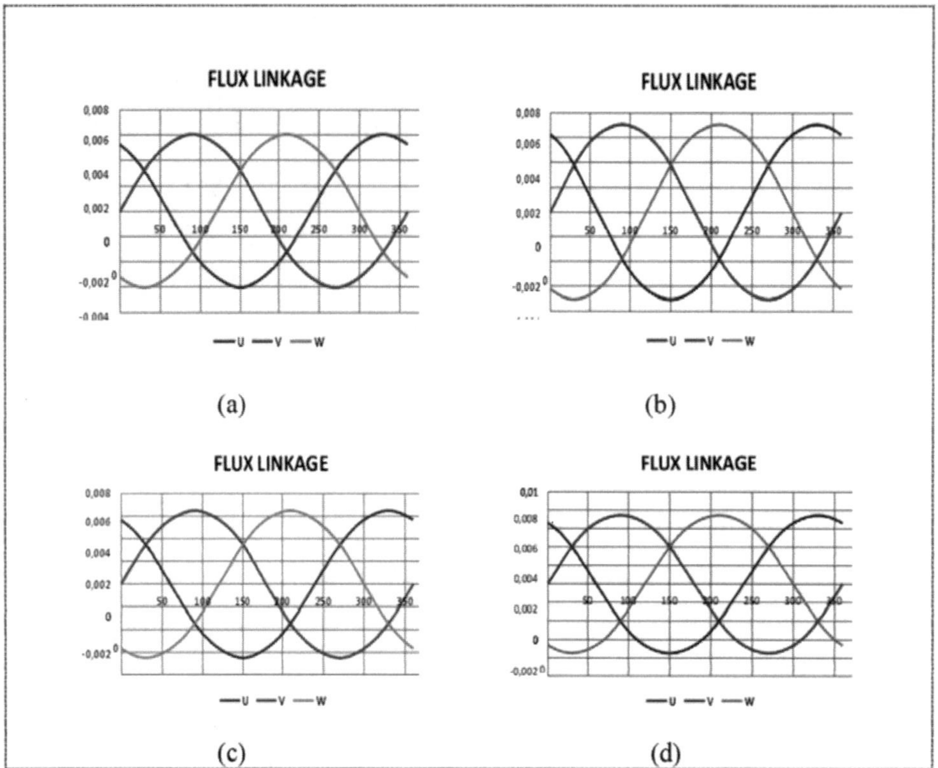

Figure 2. Graphics Flux Linkage (a)) thick 3 mm NdFeB, (b) thick 6 mm NdFeB (c) thick 3 mm, PM12 Br 1.2 Mur 1, and (d) thick 6 mm, PM12 Br 1.2 Mur 1

3.2. K_E and EMF Back Output

Once the result of the magnetic flux output is obtained, the next is calculating the EMF Back and K_E value of each phase. From the chart (d) above it can be seen that the average voltage of the modelling of these variations is at a value of 21.13122 V and the value K_E be obtained for this PMSG variation model 0.20 Vs/rad for each 360° rotor rotation. This variation model has the best average EMF Back output and K_E value among the other three models (a, b, and c). This proves that the thickness and material used on the permanent magnet will affect the resulting voltage when the rotor rotational

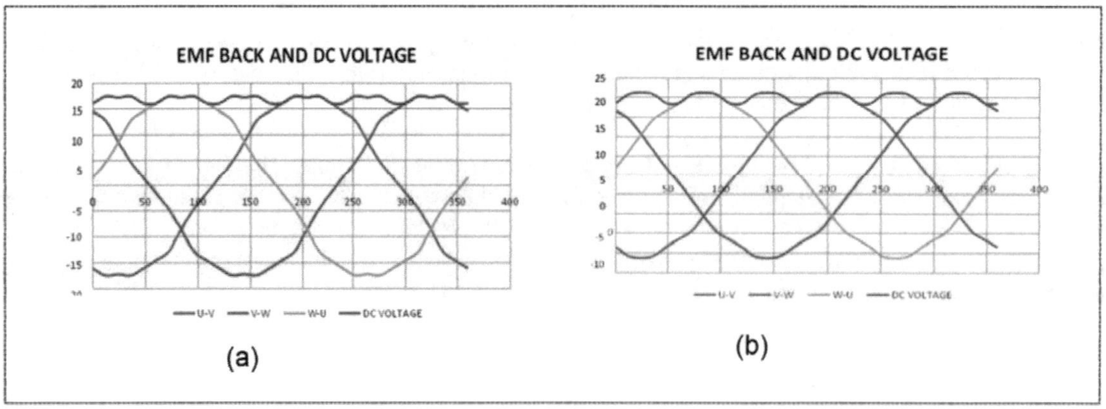

IOP Conf. Series: Materials Science and Engineering **990** (2020) 012006 doi:10.1088/1757-899X/990/1/012006

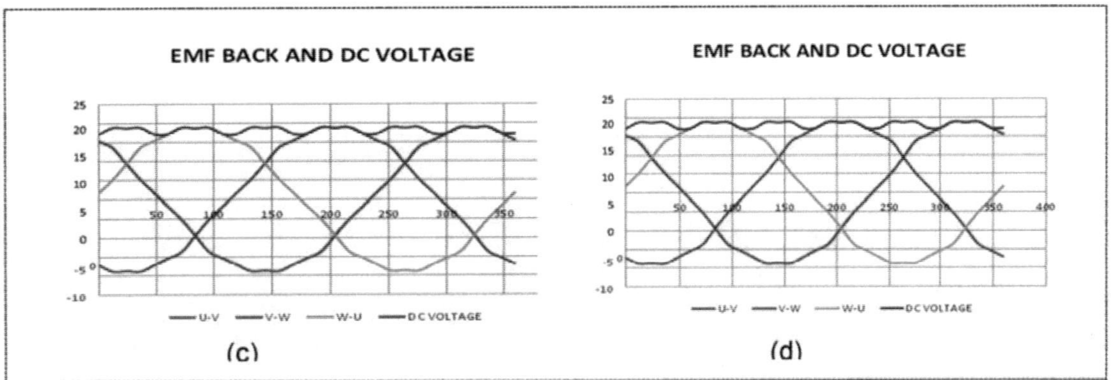

Figure 3. Graphics EMF Back and DC Voltage (a)) thick 3 mm NdFeB, (b) thick 6 mm NdFeB (c) thick 3 mm, PM12 Br 1.2 Mur 1, and (d) thick 6 mm, PM12 Br 1.2 Mur 1

4. Conclusion

The test results for four variations resulted in the conclusion that the permanent Magnet Synchronous Generator (PMSG) 12 Slot 8 Pole with FEM using Magnet 7,5 Software was variated in the permanent material magnets using PM12 Br 1.2 Mur 1 with a thickness 6 mm magnet resulted in the output of EMF Back and value K_E the largest of other variations of PMSG 12S8P. This proves that the thickness and permanent material of the magnet greatly affects the output voltage that the generator will generate. This means that the larger the dimension or thickness of the permanent magnet will lead to a wider spread of flux resulting in a better voltage. The resulting EMF Back is heavily influenced by the value of the magnetic flux.

Acknowledgment
The author thanked PT LanteraBumi Nusantara and UIN Suska Riau who have helped the authors complete this research in a timely manner.

References

[1] Piggot H 2000 *Windpower Workshop, Peninsula* (British: Wind Energy Association).
[2] Chapman S J 2005 *Electric Machinery Fundamentals* (New York: McGraw-Hill) 4th edition.
[3] Ummami, Muhammad Irsyadul 2018. *Desain Generator Sinkron Magnet PermanenJenis Neodymium Iron Boron Untuk PLTB Daya 500 Watt Menggunakan Perangkat Lunak Mag Net Infolytica* UniversitasMataram.
[4] Hanselmen, DC 2006. *Brushless Permanent Magnet Motor Design* (New York: McGraw-Hill).
[5] Prasetyo Y 2019 *Analisis Perbandingan Bahan Material Magnet Dalam Pemodelan Permanent Magnet Synchronous Generator (PMSG) 12 Slot 8 Kutub Dengan Menggunakan Finite Elemen Method (FEM) Software* UniversitasMuhammadiyah Surakarta.
[6] Nugroho W B, Kusuma I R, dan S Sarwitto S 2014 *Jurnal Teknik* **3**
[7] Arifianto I, Hasibuan M R2018 *Seminar Nasional Microwave Antena dan Propagasi (SMAP)* pp 43-47.

Design and Implementation of Ventilator for Breathing Apparatus

Hidayat[1], Saiful. J [1], Iman. S[2], Suprapto[1], Aidil. I[3], Eddy. S[1]
[1] Electro Engineering, Universitas Bung Hatta, Padang, Indonesia
[2] Mechanical Engineering, Universitas Bung Hatta, Padang Indonesia.
[3] Industrial Engineering, Universitas Bung Hatta, Padang Indonesia.

E-mail: hidayat@bunghatta.ac.id

Abstract. This paper present the design and implementation of a ventilator for a breathing apparatus. The ventilators that developed are the positive pressure ventilation type, the cycled volume mode and the controlled mechanical ventilation method (CMV). The technology that used is the ambu bag. The design and manufacture includes an automatic ambu bag pressing mechanical system and a control system using the arduino microcontroller. The mechanical design that is compact, lightweight, energy saving, low cost, meets aesthetics and is friendly makes this tool attractive so it is not scary for patients. BPM (beep per minute), IER (inspiratory expiratory ratio) and TV (tidal volume) parameters can be controlled via a touch screen or remotely using a mobile phone via 4 buttons remotely to facilitate and avoid transmission of Covid-19 to medical personnel. BPM can be set from 5-20 BPM, IER from 1: 1 to 1: 4 and TV from 40-100%. The results of ventilator testing that have been done show a good response.

1. Introduction

Ventilator as a breathing apparatus has recently become a hot topic at the national and international. The issue is the lack of availability of a ventilator to treat Covid-19 patients. Most of the deaths of corona patients in Japan can be caused by a lack of ventilators [1]. The number of Japanese-made ventilators is very lacking in Japan, so the Japanese government urges ventilator manufacturers to increase production and facilitate registration [2]. The number of positive patients infected with corona in Indonesia is more than 8,057 people and in the world 2.73 million people [3]. The main complaint of Covid-19 patients is generally respiratory problems ranging from mild to severe disorders due to lung infections which can lead to death [4]. In several cases, the corona virus can cause damage to the lungs, so that the body's oxygen levels drop and make breathing difficult. In medical science, patients who have difficulty breathing, breathing is assisted by a ventilator. A ventilator is a medical device used to help patients who have difficulty breathing or are unable to breathe on their own. The ventilator will take over the patient's breathing. Not different from other countries, one of the problems in handling corona patients in Indonesia is the limited number of ventilators available in hospitals. Besides the high price, the stock to be purchased is also not available. This is what has prompted several foreign and domestic institutions and universities to design practical, economical and ready-to-

[1] hidayat@bunghatta.ac.id.

Content from this work may be used under the terms of the Creative Commons Attribution 3.0 licence. Any further distribution of this work must maintain attribution to the author(s) and the title of the work, journal citation and DOI.
Published under licence by IOP Publishing Ltd

use ventilators. ITB with its Vent-I, ITS with Robot Ventilator, UI, UGM, UNS, BPPT, ESDM Research and Development and so on. Likewise with overseas universities such as MIT in the United States, Oxford University, Rice University, etc. Including multinational industries such as the automotive industry, PT. Pindad, PT. LEN and so on. Because of the urgency of ventilators these days, these works are published openly (open source) and researchers are allowed to develop them without the need for permission. One of them, recently Dr. Saud Anwar in America succeeded in making a Ventilator that can serve 7 patients at once. Dr. Saud Anwar gave a correspondent address for those interested in making the ventilator he found

2. The Theory of Ventilator

Mechanical ventilation is an attempt to assist breathing with a mechanical breath aid. Ventilator as a means of replacing the chest pump function that is experiencing fatigue or failure. Mechanical ventilation is used to assist or replace spontaneous breathing. Mechanical ventilation is applied with a special tool that can support the function of ventilation and improve oxygenation through the use of gases with high oxygen content and positive pressure. In its current development, mechanical breathing aids are not only a substitute for the function of the chest pump, but more broadly, namely overcoming pulmonary ventilation-perfusion disorders, so that this breathing device is agreed as a life-saving tool for critical patients who need intensive therapy. The main objectives of mechanical ventilation allowance are to ensure adequate ventilation, reduce breath work, and correct impaired oxygen exchange in the alveoli. The function of the ventilator, expanding the lungs during inspiration, can regulate the time from inspiration to expiration, prevent the lung from contracting during expiration, and can adjust the time from the expiration phase to the phase of inspiration. All sophisticated mechanical ventilators are equipped with a pressure gauge monitor, a pressure limiter to prevent the lungs from barotrauma (pressure limiting device), high and low pressure alarms, and pulmonary volume regulator (spirometer).

2. 1. Classification of the ventilator

The classification of ventilator such as negative pressure tank respiratory support (NPTRS) and positive pressure ventilation (PPV). NPTRS is the patient is placed in a cylinder that is pressurized by sub-atmospheric air (negative pressure), causing the chest to expand and the airway pressure to be negative. PPV will provide positive pressure above atmospheric pressure so that the chest and lungs expand in the inspiratory phase, then at the end of inspiration the pressure returns to equal atmospheric pressure so that air exits passively in the expiration phase. During positive pressure ventilation, pulmonary inflation is achieved by periodically applying positive pressure to the upper airway through a tight mask (non-invasive mechanical ventilation) or through an endotracheal tube or tracheostomy.

2. 2. Ventilator Model

Ventilator mode is divided based on cycling (change from inspiration to expiration) such as:

a) Pressure Limited / Pressure cycled

The pressure-cycle ventilator goes into the expiratory phase when the air pressure reaches a predetermined level. Tidal Volume (TV) and inspiratory time varied, with respect to airway and pulmonary resistance and circuit compliance. In its application, this tool is easier to accelerate by the patient's breathing effort, but when the airway resistance increases or the chest or lung stretching power increases, there will be a decrease in tidal volume and minute volume.

b) The time-cycled

Time -Cycled ventilator enters the expiratory phase after a predetermined interval calculated from the start of inspiration. TV is a product of the time of inspiration and the flow rate of inspiration. Time-cycled ventilators are commonly used for neonates and in the operating room.

c) Cycled volume

This type of ventilator can produce a certain volume that is tailored to the patient's needs. When the set volume is reached, the inspiration phase ends. Many ventilators for adult patients are volume-cycled but equipped with a secondary limit on inspiratory pressure to protect the lung from baro trauma. If the inspired pressure exceeds the pressure limit, the engine cycle continues on to expiration even if the selected volume has not been conveyed.

d) Flow cycled

The inspiratory phase changes to expiration when the air flow falls to a certain level. Flow-cycle ventilators have pressure and flow sensors that allow the ventilator to monitor inspiratory flow at a preset inspired pressure; when this flow reaches a predetermined level.

3. Design and Implementation

3. 1. Description

The ventilators developed are the Positive Pressure Ventilation type, the cycled volume mode and the Controlled Mechanical Ventilation method (CMV). The technology used is the ambu bag. The design and manufacture includes an automatic Ambu Bag pressing mechanical system and a control system using the Arduino Microcontroller. A mechanical design that is compact, lightweight, meets aesthetics and is friendly makes this tool attractive so it is not scary for patients. BPM, IER and TV parameters can be controlled via a touch screen or remotely using a cell phone via 4 buttons remotely to facilitate and avoid transmission of Covid-19 to medical personnel. BPM can be set from 5-20 BPM, IER from 1: 1 to 1: 4 and TV from 40-100%. The ambu-bag technology was chosen because it is cheaper and can be produced in mass quantities.

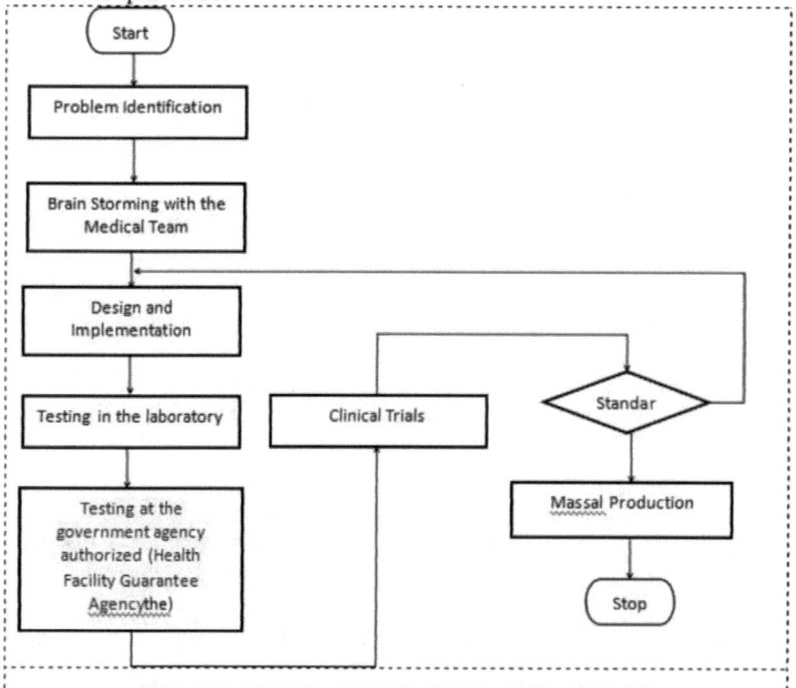

Figure 1. The step of design and implementation

The power capacity determination for the arm drive component is calculated so that it consumes minimal electrical energy. The arm shape and position of the Ambu Bag press are designed to follow the surface of the Ambu Bag, resulting in maximum tidal volume. To stabilize the position of the Ambu Bag when it is pressed, which tends to shorten, a motor that controls the position of the Ambu

Bag is installed which moves in sync with the Ambu Bag suppressor motor. The step of design and implementation is shown Fig. 1. Design specifications of ventilator such as: using the ambu bag, respiratory rate (RR) = 5-20 breaths per minute (BPM), inspiratory: expiratory time ratio (I: E ratio) = 1: 1 to 1: 4, tidal volume (TV) = volume of gas (ml) delivered during inspiration: 250-800, setting with touch screen and hand phone

3. 2. Determination of Motor Power

The mechanical system for pressing the Ambu Bag is planned to have a gripper with two motors, each of which is separated during the pressing and releasing process. So that the calculation of the power of each motor refers to the specifications of the ventilator as follows [3]. Maximum pressure on the airway (airway): P airway, max = 40 cm H_2O; Maximum respiration rate: RR max = 20 bpm; The minimum inhale / exhale ratio is 1: 4: IE ratio, min = 4; Maximum output volume: V max = 800 cm3 In the worst case, the air flow is generated at a pressure of 40 cm H2O, in 0.3 seconds.; = 60 sec / RR max: time in 1 bpm; tin hale = pressing time; tin hale = T/(1 + IE ratio); tin hale = 60 sec /RR max / (1 + IE ratio); Volumetric flow rate required in the worst conditions (peak); Q airway = V max / tin hale = 0.0027 m3 / s; The output energy (in the form of a pressurized volume flow in the airways) is: Power airway = Pair way, max Q air way = 10.46 W. However, some of the power used to compress the Ambu Bag is lost (ambu bag deformation, friction, etc.), and an estimated 50% is converted to a pressurized volume flow. Taking into account efficiency, the power required by the gripper is: Power gripper = 2 Power airway = 20.92 W. Assuming half of the motor power output is lost to mechanical and electrical inefficiencies (gears, thermal dissipation, etc.), the required power output of the motor is given by: Power motor = 2 Power gripper = 41.84 W. The ventilator is designed to use a gripper with 2 fingers. Parameters that can be obtained by measurement such as ambu bag contact area, the length of the lever arm, sweep angle. For typical prototypes, note that: Maximum Ambu Bag contact area: Abag = 90 mm x 115mm Length of the lever arm: finger = 12 cm ; Sweep angle: α_{sweep} = 30°; The maximum strength of the one finger Ambu Bag (when fully pressed) is, using the same 50% pressure transmission efficiency as before: F_{finger} = 2 A bag Pair way, max = 81,199 N. The maximum torque required for each finger is: finger = F_{finger} l_{finger} = 9.74 N.m. The power required for a two-finger gripper using the sweep angle rate (in 0.3 seconds): $P_{gripper}$ = 2 x finger w_{finger} = 34.01 W. The total power for the motor (assuming a single motor) with an additional 50% and the same gearbox efficiency, is obtained: P_{motor} = 2 x $P_{gripper}$ = 68.03 W ~ 70 W.

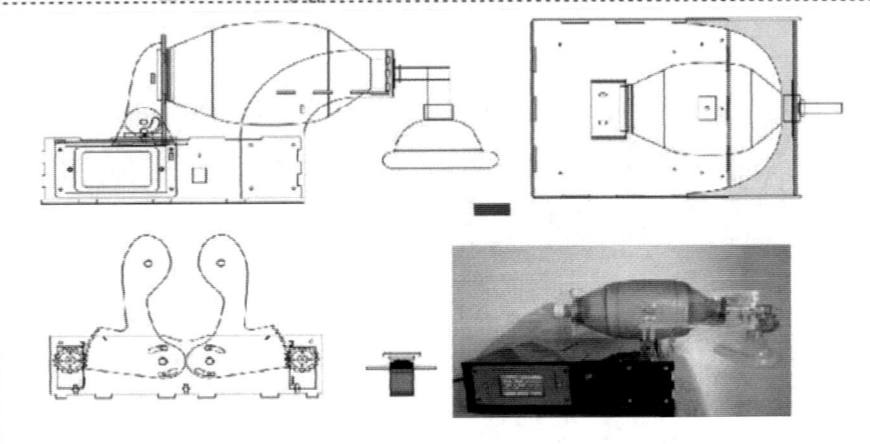

Figure 2. Ventilator Engineering Drawing.

4. Test Result and Analysis.

The test was carried out using the following equipment, that are computer, stopwatch, vessel and measuring cup. The step of testing technical are given the RR, IER and TV values, noted the time pressing (Inspiratory, Ti) releases (Expiratory, Te) the ambu bag, noted the BPM value, measured the volume based on the amount of water spilled and calculated the IER value. The first test is given setting value of BPM = 15, IER = 1: 1 and TV = 80%, then the ventilator is activated. The results obtained are BPM = 15, IER = 1: 1 and TV = 80%. This shows that the ventilator can respond according to the amount given by the ventilator. Then second test is given, namely the setting BPM = 15, IER = 1: 2 and TV = 80%, then the ventilator is activated. The results obtained are BPM = 15, IER = 1: 2 and TV = 80%. This shows that the ventilator can respond according to a given amount. By the test results, it can be seen that the ventilator is able to respond to the given setting values quickly. The test results for the other parameters can be seen in Table 1. Changes to test parameters can also be done remotely via hand pone.

Table 1. The test result of ventilator

No	Setting value			Actual Value				
	BPM	IER	TV (%)	BPM	Ti (dt)	Te(dt)	IER	TV(%)
1	15	1:1	80%	15	2	2	1:1	80%
2	15	1:2	80%	15	1,30	2,60	1:2	80%
3	12	1:2	80%	12	1,60	3,30	1:2	80%
4	10	1:2	80%	12	2	4	1:2	80%
5	15	1:1	60%	15	2	2	1:1	60%
6	15	1:2	60%	15	1,30	2,60	1:2	60%
7	12	1:2	70%	12	1,60	3,30	1:2	70%
8	10	1:2	70%	12	2	4	1:2	70%

5. Conclusion

The ventilator is designed to be very friendly in terms of performance and physical appearance is not scary Changes to test parameters can also be done remotely via hand pone. Every change in the parameters given, the ventilator can respond and work as desired. Ventilator testing will continue to the testing phase at the Medan Health Facilitation Guarantee Agency, which is an institution that provides guarantees for medical equipment, namely BPFK Medan

Reference

[1] Hariz Barak, 2020 *Japan is Overwhelmed by the Surge of New Corona COVID-19 Cases, Lack of Ventilators and ICUs*
[2] Kevin Adrian, 2019 *Ventilator Know, Benefits, and Drawbacks*
[3] Jeffrey Ebib, MD, 2020 *Global Ventilator Projects*
[4] Klein KR, Nagel NE 2007 *Mass medical evacuation: Hurricane Katrina and nursing experiences at the New Orleans airport. Disaster Manag Response* Apr-Jun;5(2):56-61
[5] Aït-Khaled N, Enarson D, Bousquet J 2001 *Chronic respiratory diseases in developing countries: the burden and strategies for prevention and management*, Bulletin of the World Health Organization, 79 (10)

[6] Abdul M.A, etc, 2010 *Design and Prototyping of a Low-cost Portable Mechanical Ventilator* Proceedings of the 2010 Design of Medical Devices Conference DMD2010 April 13-15 Minneapolis, MN, USA

IOP Conf. Series: Materials Science and Engineering **990** (2020) 012008 doi:10.1088/1757-899X/990/1/012008

Preliminary Study of Harmonic Generated by Household Appliances

I Nisja[1], S. Hardi[2], Mirzazoni[1]and Hidayat[1]
[1]Department of Electrical Engineering, Universitas Bung Hatta, Padang, Indonesia
[2] Department of Electrical Engineering, Universitas Sumatera Utara, Medan, Indonesia

E-mail: drindra765@bunghatta.ac.id

Abstract. This paper is preliminary study with focused on the harmonics generated by household appliances. There are only several household appliances that considered in this paper are television set, laptop computer and compact fluorescent lamp (CFL). To measure the input current, the measurement equipment was installed at the input side of equipment under the test. The important parameters that have been recorded are individual harmonic currents, total harmonics distortion, power factor, input current and input voltage wave form. From the measurement results, television sets, laptop computers and CFLs have been found generate the THD_i of 177%, 273.66% and 154.2% respectively. If it is widely used in residential areas, it is likely that there will be increased harmonic distortion in the electric power distribution system. Because this research is still in its early stages, at the end of this research it will be known how much harmonics is generated by the load in the residential area. If amount of harmonics generated by the housing load that exceeds the limit, so the next step of this research a harmonics filter design will carried out.

1. Introduction

Alternating current (AC) in distribution systems are designed to operate at constant frequency of 50 Hz or 60 Hz and specified voltage levels with constant magnitudes. With the development of sophisticated technology in power electronics over the last decade, the use of power electronic devices in household appliances has increased.Due to the switching operation of the power electronic equipment, they become harmonic generating devices. So, the application of power electronics equipment in household appliances has led to higher harmonic distortion for both load voltage and current waveforms. Harmonic currents generated by household appliances are individually too small to cause of any appreciable distortion in distribution feeders. However, as the numbers of household appliances use increases, the cumulative effect of these loads has the capability to elevate the harmonic distortion in distribution system [1], [2], [3], [4]. In fact, the source of small-scale harmonics is not important to study, but due to widespread use, it becomes important to study.
The major problems associated with the harmonics generated by single-phase loads are excessive neutral currents and communication interference [5]. Single phase nonlinear residential loads have always existed and traditionally included such items that use diode bridge rectifier (DBR) like: desktop computers, laptop computers, television sets, home theater etc. Harmonic effects and analysis becomes an important and necessary task in electrical power distribution system because of the fact

[1] drindra765@bunghatta.ac.id.

Content from this work may be used under the terms of the Creative Commons Attribution 3.0 licence. Any further distribution of this work must maintain attribution to the author(s) and the title of the work, journal citation and DOI.
Published under licence by IOP Publishing Ltd

that most of switching devices is widely and heavily used in single-phase power electronic loads [6], [7], [8].

2. Residential Power Electronic Loads

Generally, the small harmonic producing in household appliances can be broadly divided into two groups:

- The first group contains those household appliances that contain of single-phase capacitive-filtered diode bridge rectifier (DBR) like personal computers, laptop, color television, electronic ballasts for gas discharge lamps, battery chargers and small adjustable speed drives (ASD) for heat pumps [2], [3].
- The second load group contains the fluorescent lamps (FL) and compact fluorescent lamps (CFL) that employ magnetic ballast or electronic ballast [9], [10].

In this paper the nonlinear load appliances under consideration are television, laptop computer and compact fluorescent lamp. Currently almost all houses have television, both in cities and in rural areas. Likewise, the use of laptops has also been widely used. They raise our working efficiency, but give large amount of harmonics to the low voltage networks [6], [11], [12]. They inject enormous harmonics into the low voltage networks in every time.Single-phase capacitor DBR is most widely used for power supply in low power applications. This type of rectifier is used on computers and color televisions as a power supply. This rectifier will produce harmonics and inject them into the electric power distribution system. This harmonic distortion has several detrimental effects including overheating of distribution system components, mechanical oscillations in generators, motors, capacitors and insulation failure due to electrical resonance [13].

3. Single-Phase Bridge Rectifier Model in Time Domain

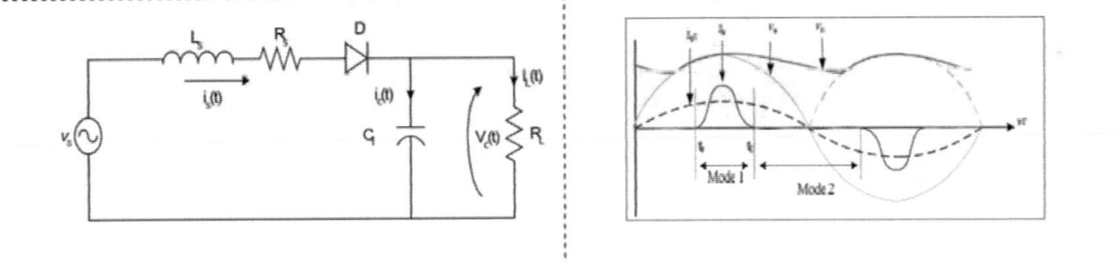

Figure 1: (a) Equivalent circuit of capacitor filtered DBR load model, (b) Supply voltage, input current, and output voltage

Figure 1(a) shows the equivalent circuit of capacitor filtered DBR load model and waveforms. The circuit operating conditions are to be assumed with highly discontinuous input current (i_s). Figure 1 (b) shows the fundamental input current, input voltage, pulsed input current and output voltage waveform generated by single phase capacitive filtered DBR [14].

As shown in Figure 1 (b), the input current (i_s) was already zero prior to zero crossing of input voltage v_s junction, for each half cycle of input waveform. There are two modes operation for this circuit. Mode 1, the diode is forward biased and the capacitor charges through the supply, the output voltage like charging voltage of capacitor. Mode 2, the diode is reversing biased and the capacitor discharges through the load, the output voltage like discharge voltage of capacitor.

The input current in steady state is the sum of its Fourier component as [15]:

$$i_s(t) = i_{s1}(t) + \sum_{n \neq 1}^{\infty} i_{sn}(t) \tag{1}$$

i_{s1} is fundamental component of input current, i_{sn} is harmonic component of input current
The Total Harmonic Distortion current (THD_i) according to IEEE definition is [16]:

$$THD = \frac{\sqrt{\sum_{n=2}^{N} I_n^2}}{I_1} \tag{2}$$

where I_1 is rms of the fundamental current waveform, I_n is rms n order harmonic of current waveform.

4. Methodology

The equipment that are using for harmonic data assemble is a power analyzer PM 100 and Visual Power Analysis software (VPAS Lite). The experiment was done individually by implemented harmonic measurement on the household appliances such as color television, laptop computer and CFL and the device gives data of harmonic order from 2 until 40 for 21".The measurement is to examine the harmonics and to determine the contribution to the system harmonic levels from the household appliances. Figure 2 shows the harmonics measurement scheme. Harmonic analysis is designed to perform the harmonic study for the household appliances. For this harmonic analysis, the results will include individual current distortion, THD_i and graphs level of harmonic current.

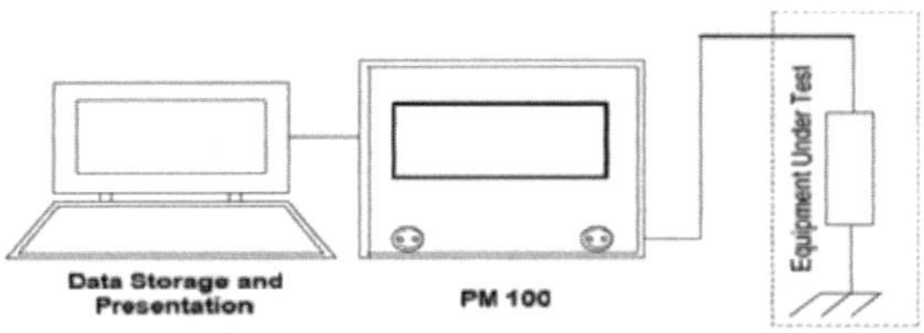

Figure 2: Harmonics measurement scheme

5. Measurement Results and Analysis

5.1 Television set

Measured waveform of input voltage, current and harmonic spectrum for a 21" television with 230 V input voltage, 67watt is shown in Figure 3(a) and (b). The voltage supply was being close to sinusoidal with THD_v = 0.317%. While the current waveform was found significantly distorted with THD_i = 177%. The 3rd harmonics until 11th harmonic order are considered high in magnitude. The odd input currents harmonic for 21' television has been considered high, but the even harmonics can be ignored due to small value. The magnitude of third harmonics current is 93.1% of the fundamental current, and the magnitude of twenty-first harmonic current is 4.4% of fundamental current. The harmonic trend is obtained rather high for low order harmonics and low harmonics magnitude for the higher order.

V_RMS (Volts)	: 230.33	I_Peak (Amps)	: 2.229	Frequency(Hz)	: 50.00
Power (Watts)	: 67.0	I_Fund (Amps)	: 0.275	Power Factor	: 0.500
Crest Factor	: 3.919	THD_i (%	: 177		

IOP Conf. Series: Materials Science and Engineering **990** (2020) 012008 doi:10.1088/1757-899X/990/1/012008

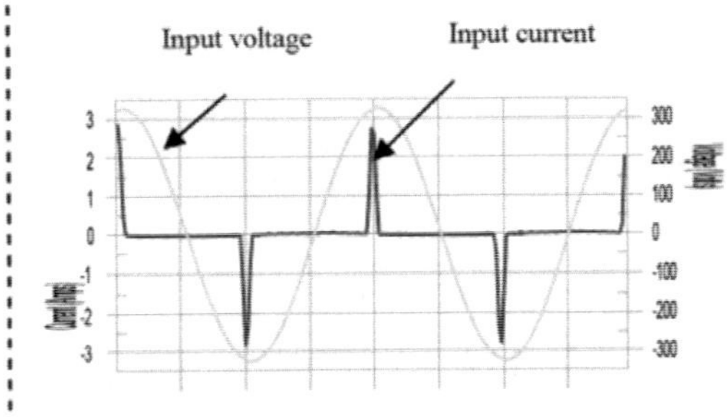

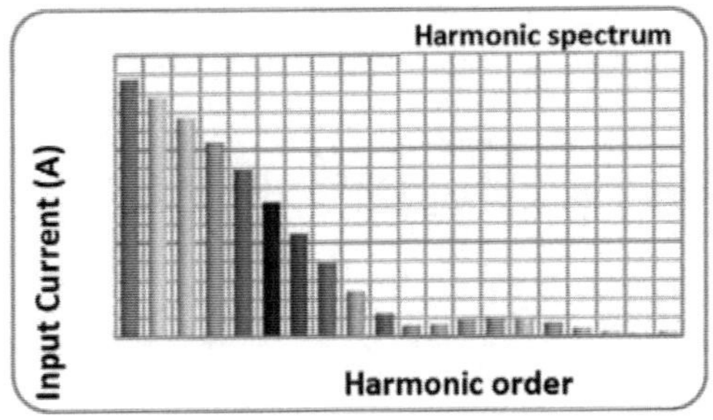

Figure 3.(a) Input voltage and current waveforms(b) Harmonic spectrum colortelevision 21"

5.2 Laptop Computer

Laptop computer has produced significant current distortion at input side. Figure 4 (a) shows the input current significantly distorted and Figure 4(b) shows harmonic spectrum input currents which content of harmonics with THD_i of 273.66% dominantly by odd harmonic. The even harmonics can be ignored due to too small value. The magnitude of second harmonic input current only 1.3% of fundamental current, but magnitude of third harmonic is 94.67% and thirty-ninth harmonic is 13.3%.

Parameter values during test:

V_RMS (Volts): 230.30	Frequency(Hz) : 50.00	I_Peak (Amps) : 1.739
I_RMS (Amps): 0.254	I_Fund (Amps): 0.075	Crest Factor : 7.089
Power (Watts) : 19.3	Power Factor : 0.330	THD_i(%) : 273.66

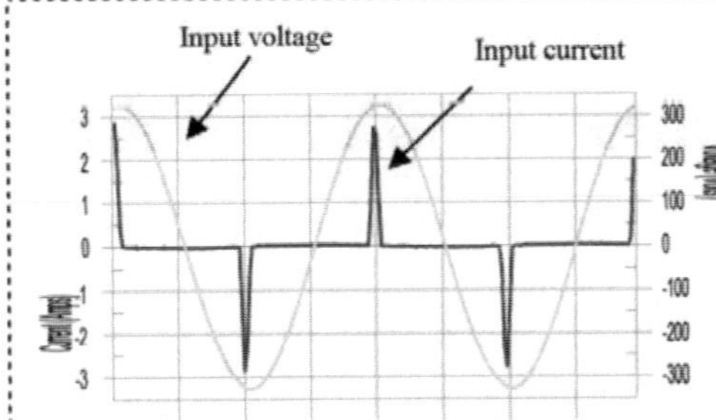

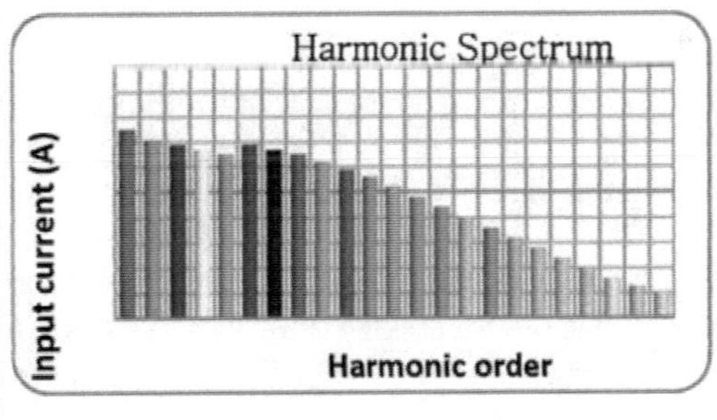

Figure 4: (a) Input voltage and current wave forms (b) Harmonic spectrum Laptop Computer

5.3 Compact Fluorescent Lamp with Electronic Ballast

Parameter values during test:

V_RMS (Volts): 230.33	Frequency(Hz): 50.00	I_Peak (Amps):0.245
I_RMS (Amps):0.059	I_Fund (Amps: 0.029	Crest Factor:4.290
Power (Watts): 7.0	Power Factor:0.530	THD_i (%): 154.2

IOP Conf. Series: Materials Science and Engineering **990** (2020) 012008 doi:10.1088/1757-899X/990/1/012008

The measurement voltage and current waveform of a 7-watt fluorescent lamp is shown in Figure 5 (a).

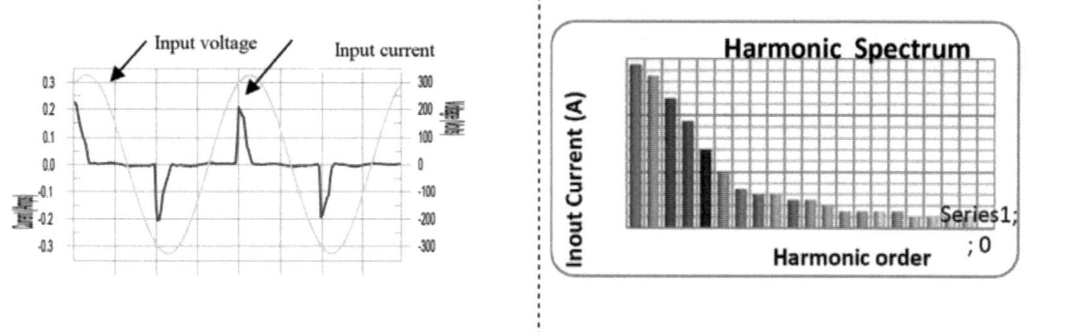

Figure 5: (a) Input voltage and current waveforms (b) Harmonic spectrum Laptop Computer

The current waveform is much distorted with THD_i = 154.2% and input voltage waveform very close to being sinusoidal with THD is 0.3 % due to low input impedance. The harmonics spectrum injected by individual fluorescent lamp with electronic ballast is shown in Figure 5 (b). It is shown that the energy saving lamp with electronic ballast injects significant harmonic distortion into the distribution system. The measurement results show that fluorescent lamps with electronic ballasts produce quite high harmonics and the large number of uses of these lamps in residential complexes will contribute to the harmonics in the electric power distribution system. This research will still be continued by measuring other household appliances such as home theatre, kitchen appliances and all kind of lighting and heating equipment. So this research will show the overall harmonics generated by household appliances.

6. Conclusion

After measuring and analyzing the harmonics produced by television sets, laptop computers and CFLs, the following conclusions can be drawn: The harmonic currents produced by the three household appliances are quite high. The television sets, laptop computers and CFLs have been found to generate THD_i of 177%, 273.66% and 154.2% respectively. The household appliances in the first group show a high source of harmonics current compared to the second group. Due to numbers of nonlinear residential loads are increasing from time to time, so that the cumulative current harmonics produced by household appliances will contribute significantly to the harmonic in electrical distribution system.

References

[1] E.F. El-Saadany 2002 *Parameters Affecting Harmonic Propagation and Distortion Levels in Non-Linear Distribution systems*Department of Electrical and Computer Engineering, University of Waterloo, Waterloo, Ontario, N2L3G1, CANADA.

[2] Juan José Mesas, Luis Sainz, Pau Sala 2015 *Statistical Study of Personal Computer Cluster Harmonic Currents from Experimental Measurement*Polytechnic University of Catalonia Barcelona, Spain.

[3] Muhammad Usman Keerio, Muhammad Shahzad Bajwa, Abdul Sattar Saand, Munwar Ayaz 2017 International Journal of Advanced Computer Science and Applications **8** 240-247.

[4] N. Indra, D. Ismail, M. Syafrudin, R. Chan Bahaudin, H. Surya 2011 International Power Engineering and Optimization Conference **5**

[5]. H.O.Aintablian and H.W.Hill, Jr. *Harmonic Currents Generated by Personal Computers and their Effects on the Distribution System Neutral Current.*

[6] Zhou Yong Zhengzhou 1997 International Conference on Advances in Power System Control, Operation and Management **4**.

[7] Angula Nashandi and Gary Atkinson-Hope 2007 *Impact of large numbers of CFLs on distribution system* Cape Peninsula University of Technology.

[8] Roy, Jinia, and Mather, Barry A. 2019 IEEE Texas Power and Energy Conference (TPEC) 1-6 Web. doi:10.1109/TPEC.2019.8662188.

[9] Andrés C. Henao-Muñoz, Javier G. Herrera-Murcia, Andrés J. Saavedra-Montes 2018 TecnoL **21**

[10] Jing Yong, Liang Chen, Alexandre B. Nassif, and Wilsun Xu 2010IEEE Transactions On Power Delivery **25**.

[11] Alexandre Nassif, Jing Yong *Harmonic Characteristics of Home Appliances Department of Electrical and Computer Engineering* University of Alberta

[12] H. Mageedl, A. S. Nada2, S. Abu-Zaid2 and R. S. Salah Eldeen 2019 Journal of Metrology Society of India 559–572.

[13] *Effect of Harmonics on Equipmen 1993Report of the IEEE Task Force on Harmonic Impacts* IEEE Trans. PWRD **8** 672-680.

[14] D.O. Koval and C. Carter 1997 *Power Quality Characteristics of Computer Loads* IEEE Trans. IA **33** 613-621.

[15] Mohan Ned, Underlan, T. M., Robbins, W. P. 2003 *Power Electronics, Converter, Applications and Design* Third Edition John Wiley and Sons Inc.

[16] Rashid, M. H. 2004 *Power Electronics, Circuits, Devices and Applications* Third Edition Pearson Education, Inc.

IOP Conf. Series: Materials Science and Engineering **990** (2020) 012009 doi:10.1088/1757-899X/990/1/012009

Productivity Improvement Through Innovation of Production Facilities in MSMEs

D Mufti[1], Y Mahjoedin[2], A Iksan[3]
[1,3]Department of Industrial Engineering, Universitas Bung Hatta, Padang, Indonesia
[2]Department of Mechanical Engineering, Universitas Bung Hatta, Padang, Indonesia

E-mail: dessimufti@bunghatta.ac.id

Abstract. Sakura cake is a traditional food from District Cangkiang, Agam Regency, West Sumatra. This cake has been marketed to Bukittingi, Padang and has been used as a typical food from West Sumatra. In Cangkiang there are a number of similar Sakura cake micro-businesses and still use coconut shell as a moulding cake. However, the production capacity with coconut shell is still low even unable to meet demand. In addition, the use of coconut shell can risk injury to workers. So an innovation is made on the tools that are already on the market and can be used as a substitute for it. In research on productivity calculation using the Objective Matrix (OMAX) method by using production data for six periods. The productivity criteria measured are the use of raw materials, the use of working hours, fuel consumption and electricity use. The results of this study found that achieving the highest score was found in the criteria for use of working hours, at level 9, which was reached at the end of the survey period.

1. Introduction

One of the potential for micro, small and medium enterprises is to be able to open employment opportunities and improve people's lives better [1]. The Small Scale Industry sector plays an important role in the development of industry in any country. The Importance of the Small-Scale Industry Sector is recognized throughout the world because of its significant contribution to satisfying various socioeconomic objectives, such as higher employment growth, output, export promotion and entrepreneurship building [2]. Micro, Small and Medium Enterprises (MSMEs) are productive businesses owned by individuals or individual business entities that have annual sales of between 300 million and 50 billion rupiahs [3]. These small, micro, small and medium enterprises (MSMEs) also contribute to the increase in gross domestic product [4].

The MSME sector is a populist economic activity in which the products produced are usually in the form of daily necessities needed by the community [5]. MSMEs become the backbone of the populist economic system to reduce the problem of poverty and its development is able to expand the economic base and can make a significant contribution in improving the regional economy and national economic resilience [6]. For this reason, it is necessary to increase the productivity of MSMEs that aim to keep these MSMEs through various supporting fields. Increased productivity can be done from all directions. As an example of 5S implementation for area preparation, in the end it can compile an effective and efficient work area so that it minimizes time as one of waste. Another example can also make improvements to work methods that will also minimize work risks, work activities, defects which will

[1] dessimufti@bunghatta.ac.id

Content from this work may be used under the terms of the Creative Commons Attribution 3.0 licence. Any further distribution of this work must maintain attribution to the author(s) and the title of the work, journal citation and DOI.
Published under licence by IOP Publishing Ltd

IOP Conf. Series: Materials Science and Engineering **990** (2020) 012009 doi:10.1088/1757-899X/990/1/012009

later reduce production costs. One of the disadvantages and needs to be corrected from this micro business is the work facilities or technology used are not ergonomic [7]. Using low process technology will make it difficult for MSMEs to increase productivity [8]. The thing that becomes the main driver in the source of increasing productivity besides the scale of efficiency is technological change [9]. SMEs in Indonesia consider technology investments important to increase productivity and cost efficiency will be better with the use of technology [10]. Technology improvements in SMEs for snacks are needed so as not to be left behind by current technological developments [11].

Cangkiang is one of the agricultural areas in Agam. Cangkiang is famous for making Sakura cakes. This cake is a traditional cake that must exist when there are traditional events such as weddings and so on. At present the market has reached West Sumatra and has entered into modern markets or souvenir shops [12]. The equipment used to make sakura cakes is "batok kelapa" which is perforated as a tool for moulding. Batok kelapa which is currently still being used with consideration and maintaining this simple technology is to maintain the distinctive shape of the product produced. One of the production processes for making sakura cakes is the stage of moulding it. This process takes a long time, which is around 8 (eight) hours. Currently MSMEs have replaced batok kelapa with existing technology, namely innovation from meat grinder [13]. In this study, we will observe the production that can be produced by the Sakura cake moulding using the innovative technology of meat grinder in the mould section. Conduct performance testing by comparing the time of printing the cherry cake manually using a printer that has been made. The use of this technology will be able to increase the productivity of MSMEs.

One method that can be used in productivity measurement is Objective Matrix (OMAX). The OMAX method is a partial productivity analysis developed to monitor productivity in each part of the company. The OMAX method measures productivity by objectively assessing the performance of each part of the company while looking for factors that cause a decrease in productivity.

The tools used in data retrieval are stationery, camcorder videos, stopwatches and observation sheets. Primary data collection is obtained by direct observation and recording in the field and interviews with MSME owners. The production data used in this study is data for 6 months starting from May 2019 to October 2019. This OMAX productivity calculation uses tools that have been innovated. The old printing tool is to use coconut shell as shown in Figure 1 and the printing tool used now is as shown in Figure 2.

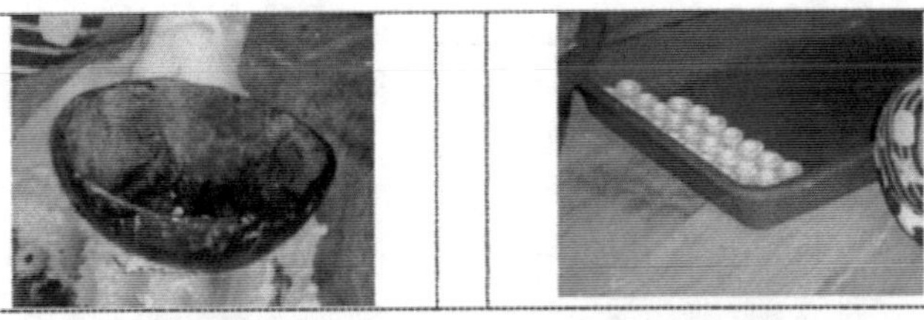

Figure 1. Existing moulding and result

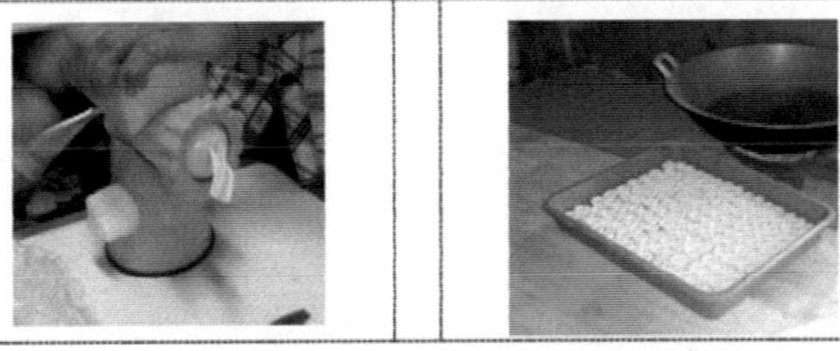

Figure 2. New Moulding and result

2. Methodology

The productivity criteria to be measured are:

(a) Criteria I is productivity of raw material usage;
(b) Criteria II is the productivity of working hours (man hour.);
(c) Criteria III is the productivity of fuel consumption and
(d) Criteria IV is productivity of electricity consumption
The criteria for measuring productivity used are in accordance with the conditions of MSME actors.
Calculate performance ratio

Performance is the level of productivity which is the ratio of each criterion per measurement period. Performance values are obtained by dividing the ratio of input to output in each criterion. The value of the performance of each criterion can be seen in Table 1 which shows the value of fluctuating performance.

Table 1. Value Of Performance Each Criteria

Period	**Performance Productivity Criteria**			
	Raw material usage	**Man hour consumption**	**Use of fuel**	**Electrical Energy**
Mei	1022	2242	9276	21455
Juni	1073	1896	9549	21711
Juli	1106	1766	10091	24811
Agustus	1081	1698	9128	22187
September	1105	1777	9262	21058
Oktober	1133	2489	5052	18504

Determine the average productivity value (level 3) Level 3 value (μ) is obtained from the average value of company performance during the measurement period. Level 3 values obtained for each productivity criterion can be seen in Table 2 whose values are obtained using a formula:

$$\mu = \frac{1}{n} + \sum_{n=1}^{n} xi \tag{1}$$

Description: μ = Average ratio of each criterion measured

n= Amount of data
xi = i-criteria ratio

Determine the highest value of productivity (level 10)

Every company must have a target to be achieved in a certain time according to the company's ability. The company target is described at level 10. Level 10 can be seen in Table 2 obtained by using the formula:

$$BKA = \mu + k \cdot \sigma \tag{2}$$

$$\sigma = \sqrt{\sum \frac{(xi-\mu)^2}{n}} \tag{3}$$

$$DA = (\sigma/\mu) \times 100\% \tag{4}$$

$$CL = 100\% \times DA \tag{5}$$

Description: BKA = Limit of Control over DA = degree of accuracy CL = confident level μ = Average ratio of each criterion n = Number of data σ = Standard Deviation k = Constants

Determine the lowest productivity value (level 0)

The lowest productivity value is a value that must be avoided by companies because this value is the worst achievement. This level 0 is obtained by using the Lower Control Limit (BKB). The lowest score is in Table 2 obtained by using the formula:

$$BKB = \mu - k \,.\, \sigma \tag{6}$$

Determine the value of actual productivity (levels 1-2 and 4-9).
The actual productivity value is a possible value before the final goal. Levels 1-2 and 4-9 on each criterion can be seen in Table 3 obtained by using the formula:

$$Interval\ (1-2) = \frac{level\ 3 - level\ 0}{(3-0)} \tag{7}$$

$$Interval\ (4-9) = \frac{level\ 10 - level\ 3}{(10-3)} \tag{8}$$

Determine score, weight, and value, calculate performance indicators.

Scores are levels that indicate the value of productivity (performance) at the time of measurement. Each criterion has a different level of importance for increasing productivity. Therefore, it is necessary to weight on each criterion. In this study, weighting is done by pairwise comparison method.
Value is the result of multiplication between scores and weights on each criterion measured. The sum of value values of all criteria is used to determine the total productivity value of the company. After the score, weight, and value values obtained, the next step is to determine the value of performance indicators. Calculation of performance indicators consists of three, namely: (a) Current, which is the result of current period productivity measurements obtained from the sum of the value of each criterion; (b) Previous, namely the results of previous productivity measurements; and (c) Index, which is an indication of productivity changes that occur in the company. The index productivity (IP) value is obtained by the formula:

$$IP = ((Current\text{-}Previous)/Precious) \times 100\% \tag{9}$$

3. Result and discussion

Table 2. Determination of Average Values (Level 3 Or M

No	Productivity Criteria	Level (3)	Level(10)	Level(0)
1	Raw material usage	1004	1322	802
2	Man hour consumption	1986	2535	1438
3	Use of fuel	8870	10610	7131
4	Electrical Energy	20148	26871	1454

Table 3. Levels 1-2 and 4-9 for Each Criteria

Raw material usage	Raw material usage	Use of fuel	Electrical Energy	Level
1178	2458	10361	25956	9

1148	2379	10112	25040	8
1120	2300	9864	24124	7
1091	222	9616	23209	6
1062	2144	9367	22294	5
1032	2065	9118	21378	4
1004	1986	8870	20148	3
936	1804	8290	18324	2
869	1621	7710	16188	1

Determination of Score, Weight and Value

Based on the results of questionnaire data processing, the value of Consistency Ratio (CR) ≤ 0.1 was obtained, which was 0.019 so that the results of the Consistency Ratio (CR) were consistent. The weight for each criterion is shown in Table 4. The criteria that have the highest weight are the criteria for raw materials. According to respondents, raw materials are very influential in the production process to produce cake products because raw materials can affect the number of products (output) and quality produced. Criteria that has the lowest weighting value is the fuel criterion. This is because the fuel used is always available and almost never experiences interference so the level of importance is smaller than the other criteria.

Table 4. Weighting Results for Each Productivity Criteria

No	**Productivity Criteria**	Bobot	%
1	Raw material usage	0.446	44.6
2.	Man hour consumption	0.333	33.3
3.	Use of fuel	0.070	7.0
4.	Electrical Energy	0.151	15.1
	Total	1	100

Partial productivity evaluation is based on achieving the productivity level of each criterion. Each criterion has different influences in achieving productivity. These changes can be evaluated through a level that shows the level of productivity achieved in each measurement period. The level of achievement of productivity can be seen in Table 5.

In the raw material usage criteria, achievement of levels in October was high because the level of efficiency of raw material use was better in October, while the lowest level that occurred in September and October was due to improper processing and incompatibility of raw materials received by the company. On the criteria for use of working hours (man hour), the level of productivity that is below the average level is due to lack of motivation and supervision of the workforce in the processing process so that the company cannot reach the predetermined target. On the criteria for fuel use and electricity, achievement of productivity levels continues to decline until October with a score of 0. The level of productivity decreases to below average due to inaccuracies in determining the fuel to be used for the production process so that more attention and repairs are needed so that the productivity achieved can be even better. In October, there was a very high increase in productivity. This is because the level of the criteria for the productivity of raw materials, man hours of the production process runs smoothly, there are no obstacles in terms of receipt of raw materials and processing so that the company is able to produce cakes more than demand.

Table 5. Achievement Levels of Productivity

Level	Performance Productivity Criteria			
	Raw material usage	**Raw material usage**	**Use of fuel**	**Level**
Mei	3	8	4	4
Juni	5	2	5	4
Juli	6	1	7	7
Agustus	5	1	4	4
September	6	1	4	3
Oktober	7	9	7	2

This total productivity evaluation is carried out to determine the level of total productivity achieved by the company. Evaluation is also done by looking at the productivity index value on performance indicators in the OMAX. Identification of problems is done by direct interviews with the part of workers or MSMEs actors. The results of the interview obtained information that there are 2 factors that become productivity problems, namely material, human, machine and environmental factors. This is an internal factor of MSMEs, while external factors are influenced by funding, entrepreneurship training, business assistance and partnerships [14]. In material factors, the cause of low productivity is the raw material received is not in accordance with the company's demand and the quality of raw materials of sticky rice is not good sometimes the process of rice milling is not perfect. The printer tool causes low productivity because it is still manual. Another factor that results in low productivity is the environment. Environmental inconveniences that occur in the production sector are equated with the kitchen where SMEs are cooking. Productivity improvements are proposed after knowing the productivity achieved by the company and after identifying productivity problems. Improvements are made based on the average achievement of productivity in the period January-June 2019. Data on proposed improvements can be seen in Table 6. Waste occurs on all criteria. In the criteria of raw materials, waste is caused by improper processing so that it can cause damage to raw materials, many raw materials that fall during the printing process, as well as raw materials attached to the production machine.

Table 6. Productivity Improvements

No	Productivity Criteria	Performance before improvement	Performance after improvement
1	Raw material usage	1004	1567
2.	Man hour consumption	1986	2005
3.	Use of fuel	8870	8976
4.	Electrical Energy	20148	21567

4. Conclusion

Productivity achievements in the 2019 period fluctuated greatly. The highest score achievement is in the criteria for using working hours, namely 9, which was achieved in October. The highest total productivity index level achieved during the measurement in the 2019 period was in October at 84.69%.

Acknowledgment
This research is supported by internal research from Universitas Bung Hatta in the 2020 beginner lecturer research scheme.

References

[1] Tambunan, Tulus 2006 *Development Of Small-And Medium-Scale Industry Clusters In Indonesia, Kadin Indonesia-Jetro*

[2] Vishwa Vidyapeetham, Amrita 2015 International Journal of Mechanical And Production Engineering **3** 113-119

[3] Undang-Undang Republik Indonesia 2008 *Undang-Undang Republik Indonesia Nomor 20 Tahun 2008 Tentang Usaha Mikro Kecil dan Menengah*. Jakarta: Republik Indonesia

[4] K. P. R. Indonesia 2017 *Kontribusi UMKM Naik The contribution of SMEs increased* Jakarta.

[5] Anggraini, D 2013 Jurnal Ekonomi dan Keuangan **1** 105-116

[6] Setyanto, A. R., Samodra, B. R., & Pratama, Y. P. 2015 *Kajian Strategi Pemberdayaan UMKM Dalam Menghadapi Perdagangan Bebas Kawasan ASEAN*. Etikonomi 205-220 www.kadin-indonesia.or.id.

[7] Meilani, Difana, Hilma Raimona Zadry, Dina Rahmayanti, Dendi Adi Saputra 2018 International Journal on Advanced Science Engineering Information Technology **8** 1077-1084

[8] M. S. A. Chowdhury, M. K. G. Azam, and S. Islam 2015 Asian Business Review **2** 51–58.

[9] Dhwani Gambhir, Seema Sharma 2015 Journal of Economic and Administrative Sciences **31** 71-85, https://doi.org/10.1108/JEAS-09-2014-0021

[10] Paul Kanv 2018 *Country Head Business Banking PT Bank UOB Indonesia (UOB Indonesia)*, https://www.cnbcindonesia.com/news

[11] Hasibuan, Sawarni 2015 International Journal on Advanced Science Engineering Information Technology **6** 410-416

[12] Mufti, Dessi,.,Aidil Ikhsan, Inna Kholidasari, Yesmizarti Muchtiar, Irvandy saputra 2015 Journal Industrial Servicess Industrial Engineering Advanced Research and Application.

[13] Mufti, Dessi., Aidil Ikhsan, tri Marta Putri 2017 *International Conference on QIR (Quality in Research)* **15**

[14] Yanah, Haulah Nakhwatunnisa, Tri Amalia Sukarno 2018 Jurnal Ekonomi dan Kebijakan **11** 138-150

The Heritage Tourism Development As the Sustainable Development Goal of the Enclave Settlement: A Preliminary Research

E Purwanto[1],R Sjarief[1], ADawan[1], H Tannady[1]
[1]Department of Management& Center for Urban Studies, Universitas Pembangunan Jaya, South Tangerang, Indonesia

E-mail: edi.purwanto@upj.ac.id

Abstract. The specific target of this research is to map the values of cultural and spiritual heritage in the enclave village, Lengkong Kyai. This research will use a qualitative case study method. Data will be collected through in-depth interviews, Focus Group Discussion (FDG), related documents, and visual interpretation. The data collected will be carried out by a three-stage triangulation process, which is the primary key in collecting qualitative data. They are data triangulation, data analysis triangulation, and triangulation by comparing it with previous studies. The preliminary results are there heritage tourism potential that is developed in the enclave village. The future research will find a branding strategy that allows the potential for cultural and spiritual-based tourism in Lengkong Kyai village to be developed and its impact on the welfare of the Lengkong Kyai settlers.

1. Introduction

The urban population explosion and rapid urbanization are faster than the government's ability to provide housing [1]. The rate of population growth and the city's density encourages the population movement from the downtown city to its surrounding areas so that new cities are formed around big cities—likewise, the Jakarta case. Jakarta's population density has encouraged residents to move to surrounding cities, such as Bogor, Depok, Tangerang, and Bekasi. The high flow of urbanization and population migration is also an opportunity for real estate companies to provide settlements around Jakarta by taking advantage of the problem of housing needs in Jakarta. The presence of real estate gave birth to new cities in Jabodetabek, such as Tangerang. Apart from urban intensification and extensification policies, the development of new cities is another way out of the high population density of large cities [2].

New problems arose along with the construction of new cities and planned settlements. As indigenous people who have lived the villages for generations, they have to be evicted after agreeing to the land and building prices offered by the developers. However, many indigenous settler communities defend their lands and settlements for various reasons. The reason for the purchase price that the natives cannot accept or for other reasons is that the settlers did not give up their land or settlement to be purchased and converted into a new city because their predecessors' historical legacy is stored. These villages later became enclave settlements surrounded by elite modern urban-style housing [2].

---

[1] edi.purwanto@upj.ac.id

Content from this work may be used under the terms of the Creative Commons Attribution 3.0 licence. Any further distribution of this work must maintain attribution to the author(s) and the title of the work, journal citation and DOI.
Published under licence by IOP Publishing Ltd

LengkongKyai is one of the settlements sandwiched between luxury housing in the BSD City area. It is said that this village is believed to have existed since 1628. Prince Aria Wangsakara founded it from Sumedang, who was tasked with resolving disputes between the Mataram kingdom and the Banten kingdom. Because of his services, Prince Aria Wangsakara was given an area that was later known as the Lengkong Kyai area as his territory. Prince Aria Wangsakara is assisted by two younger siblings, namely Arya Santika and Arya Yudanegara. Lengkong Village then became the starting point for the growth of the Tangerang people's patriotic spirit under Prince Aria Wangsakara against the Dutch colonialists [3].

LengkongKyai or Lengkong Ulama is located in Hamlet 01 (RW 01), Lengkong Kulon Village, Pagedangan District, Tangerang Regency, Banten. According to the profile provided by Medtry[3], the total population of Lengkong Ulama Village is 1500 people, which are divided into three Neighbourhood (RT), namely the total population of First Neighbourhoodof 608 people and occupies an area of 4.28 hectares. The total population of Second Neighbourhood is 448 people, with an area of 4.99 hectares. Then the population in Third Neighbourhood is 444 people with an area of 1.52 hectares. In the eastern part of the village, it is bordered by Cisadane River, while in the West and North, it is bordered by BSD City housing walls, while on the south side is BSD Boulevard Utara Street.

Medtry et al. [3] researched Lengkong village and saw the village's potential as a tourist village. The focus of the suggestions and recommendations of Medtry et al. [3] are river tourism and culinary delights of the archipelago along the Cisadane riverbank. However, we see that historical heritage is the wealth of Lengkong Kyai, which has great potential for the welfare of the local community in particular. This historical heritage can have the potential for cultural and religious tourism for Lengkong Kyai.

2. Theoretical Background

Ben [4] describes three attractions of tourist visits, namely: (1) the attraction of natural beauty tourism, (2) the attractiveness of cultural tourism, and (3) the attraction of spiritual tourism. Research recommendations of Medtry et al. [3] include the Lengkong tourism potential based on the attraction of natural beauty (river tourism). Even though it is possible, two other attractions, cultural tourism and spiritual tourism have great potential to be developed for the welfare of the Lengkong Kyai community. That is why this research will investigate and explore the possibility of cultural and spiritual tourism in particular, while still considering the potential for tourism in natural beauty (rivers).

Ben [4] said that the fastest growing sector in the tourism sector is visiting cultural and historical heritage tours, especially in developing countries. Music, dances, language, religion, culinary, traditional arts, festivals, and physical, cultural heritage are forms of cultural tourism objects. The daily life of indigenous rural communities can also be a cultural tourism object.

According to Ben [4], religious tourism is a form of heritage tourism in developing countries. Pilgrimage can be referred to as a form of heritage tourism from three perspectives: visited sites (churches, mosques, temples, sacred mountains, and caves), historic routes that are sources of historical heritage (Holyland tourism), rituals or worship performed by these sites. Sites that are considered sacred are also cultural heritage. What distinguishes cultural and spiritual tourism is if cultural tourism is related to religion, the emphasis is on visits to sites, historical routes, and religious rituals. However, spiritual tourism means that tourists visit holy places to pray or worship and believe that they are closer to God.

Spiritual tourism is one of the dynamically developing areas in the tourism sector. Religious tourism is generally the subject of research in the business and social areas [5]. Norman [6], [7] identifies spiritual tourism experiences as follows: spiritual tourism as healing, seeking meaning, retreat – *muhasabah*, togetherness, and experience. And nowadays, religion is one of the topics that are often researched in the tourism industry, as well as the opinion of several researchers that tourism and religion are closely related, and religion is the main reason for the spiritual tourism movement [8].

Ben [4] said that tourist attractions with sacred values are the main attractions for tourist visits. Awareness of the same religious belief and identity is a stimulus for people to visit spiritual tourism. Examples of spiritual tourism objects in Java include mosques and tombs of the Wali Songo, temples, and other holy places, which are tourist destinations that can produce spiritual experiences for adherents. This spiritual tourism is referred to as tourism with sacred and secular values because its purpose is worship as well as tourism, inner tranquility, and joy at the same time.

3. Method

This research uses a qualitative approach. According to Yin [9], there are three principles of qualitative research data collection, namely: 1. Proof by using many sources; 2. Create a database, and 3. Build a chain of evidence. The point of evidence using multiple sources is that qualitative research relies on triangulation, namely the collection of data from various sources, all of which are relevant to the phenomenon being studied [10]. The multiple sources referred to include documentation, interviews, participant observation, field observations, and material/recording archives, and physical artefacts [9]. In this study, data collection will be carried out in several ways: in-depth interviews with key informants, Focus Group Discussion (FGD) with the community, documents, and visuals. As preliminary research, this paper is a result of the early step, namely field observation, physical artifacts, and unstructured interviews with one of the settlement leaders.

4. Preliminary Result

In this enclave village, there are historical sites, namely the Prince Aria Wangsakara Tomb and the Mosque. Prince Aria Wangsakara is a descendant of the king of Sumedang Larang, namely Sultan Syarif Abdulrohman. He is an *Ulama* who spread Islam religion and a pioneer of State Defense [11].

Around 1640 there was an attack on the *Pesantren* in Grendeng (Islamic Boarding School) by the Dutch. *Pesantren* in Grendeng was located west of the Cisadane River, now in Karawaci District, Tangerang City, which was located right across from the Dutch fort on the east of the river. The students who were knocked back then moved to a new residence in Lengkong and then built a new mosque and *Pesantren* under the leadership of Aria Wangsakara. With the blessing of the Sultan of Banten, Sultan Maulana Yusuf, Aria Wangsakara built a fort at Lengkong Kyai. Besides being known as an Islamic scholar (*Ulama*), Aria Wangsakara is also a combatant against the Dutch. The spirit of Aria Wangsakara has been passed down from generation to generation by the people of LengkongKyai[11].

The three princes died as patriots. Aria Santika died in war at Kebon Besar in 1717. He was buried in Batu Ceper District (Kramat Asem). Yuda Negara died in Cikokol in 1718. He was buried in Sangiang, Jatiuwung District. Meanwhile, Aria Wangsakara died in Ciledug in 1720. He was buried in Lengkong Kulon or Lengkong Kyai. The tomb of Aria Wangsakara is surrounded by the graves of many *Ulama*or*Kyai* who was influential in the spread of Islam in the Tangerang area.

According to local leaders, most of the villagers are descendants of the founder of Lengkong Kyai, namely the descendants of Prince Aria Wangsakara, Prince Arya Santika, and Prince Arya Yudanegara. Thus there is a very close connection between the site of the Prince Aria Wangsakara's Tomb and the residents of Lengkong Kyai, who are descendants of these leaders. Thus, in terms of the cultural context, the existence and identity of the Lengkong Kyai residents can become the spirit of cultural and religious tourism at the Prince Aria Wangsakara's Tomb.

The connection between Prince Aria Wangsakara's Tomb and Lengkong residents' identity can be a unique attraction for heritage tourism in itself. Reflecting on Iranian tourism, most importantly, based on socio-cultural dimensions of society, in the form of cultural heritage and religious attractions [12], LengkongUlama also has socio-cultural dimensions which are a tourist attraction. The Lengkong "*Palang Pintu*" ("Doorstop") culture is one of the attractions that can be seen on Youtube. Suppose the shrines of Shia figures such as Imam Reza in Mashhad and Fatima in Qom are the main tourist attractions for Iran's central heritage [12]. In that case, the tomb of Prince Aria Wangsakara in

Lengkong Ulama can also be an attraction for heritage tourism, especially for people who admire the figure of Prince Aria Wangsakara as *Ulama* and hero of independence of Indonesia.

According to Tubagus Nurfadil as the Deputy for History Division of the *Balai Adat Keariaan Tangerang*, the Vice President of Indonesia, KH. Maruf Amin is still a descendant of Prince Aria Wangsakara, both from his father's and mother's lineage [13]. Thus, the popularity of KH. Maruf Amin, as an Ulama and a Vice President, can be the basis for popularizing the village of Lengkong Ulama at the national level.

5. Conclusion and Recommendation for Future Research

The preliminary result shows that Lengkong Ulama village has the potential for heritage or spiritual tourism development. The relationship between Prince Aria Wangsakara's Tomb and his descents will be heritage tourism attraction. This heritage and spiritual tourism development by involving the enclave settlers will raise their welfare.

The limitation of this study is that it is the result of preliminary research. Therefore, a more in-depth study will be carried out to (1) map the cultural and spiritual heritage values in LengkongKyai. (2) To find out what are the hindering factors so that the potential for cultural and religious tourism is barely widely broadcast. (3) Finding a branding strategy that allows the potential for cultural and spiritual-based tourism in Lengkong Kyai Village to develop and impact the Lengkong Kyai community's welfare.

References

[1] Adams, C 1993 Pertambahan Pendudukan dan Penyerbuan Daerah Kota, dalam *Kemiskinan di Perkotaan: Bacaan untuk Antropologi Perkotaan* Editor: Parsudi Suparlan. Jakarta: Yayasan Obor Indonesia.

[2] Ischak, M., Setioko, B., and Nurgandarum, D 2017 IOP Conf. Series: Earth and Environmental Science **99** 1-13.

[3] Medtry, Haryo, B., and Madjid, K 2016 Prosiding Seminar NasionalPerencanaan Pembangunan InklusifDesa - Kota, UniversitasAndalas.

[4] Ben, S. M. (2018). *Filsafat Pariwisata: Sebuah Kajian Filsafat Praktis.* Yogyakarta: PustakaPelajar.

[5] Kaur, G. 2016Journal of Relationship Marketing 15326-343.

[6] Norman, A2012 *Literature and Aesthetics***22**20–32.

[7] Gill, C., Packer, J. &Ballantyne, R 2018 Tourism Recreation Research**43** 235-249

[8] Küpeli, T. S., Koc, B. & Hassan, A2018 Anatolia: An International Journal of Tourism and Hospitality Research**29**252-266.

[9] Yin, R. K. (2011). *Qualitative research from start to finish*. New York, NY: Guilford Press.

[10] Locke, K. D. (2011). *Grounded Theory in Management Research*. London: SAGE Publications.

[11] Ramdzy, A. (2013). Aria Wangsakara, Penyebar Agama danPeloporBela Negara, *Hits Banten,* retrieved from https://bantenhits.com/2013/06/02/aria-wangsakara-penyebar-agama-dan-pelopor-bela-negara/ on August 10, 2020.

[12] Chianeh, R. H., Chiappa, G. D. &Ghasemi, V 2018 Anatolia: An International Journal of Tourism and Hospitality Research**29**204-214.

[13] Deniansyah, R. (2020). Wapres Ma'ruf Amin Ziarah ke Makam Raden Aria Wangsakara di Tangerang, *Tangerang News.* Retrieved from http://tangerangnews.com/tangsel/read/32293/Wapres-Maruf-Amin-Ziarah-ke-Makam-Raden-Aria-Wangsakara-di-Tangerang

IOP Conf. Series: Materials Science and Engineering **990** (2020) 012011 doi:10.1088/1757-899X/990/1/012011

Combination of Oil Palm Empty Fruit Bunch and Multi Media Layer Coir Filter to Treat Water in Mendalo Darat, Jambi

R Safita[1], F Kurniawan[2], Deliza[3]
[1,2]Tarbiyah and Teacher Training Faculty, UIN Sulthan Thaha Saifuddin, Jambi, Indonesia,
[3] Chemistry Department, Faculty of Science and Technology, UIN Sulthan Thaha Saifuddin, Jambi, Indonesia,

Email: Deliza@uinjambi.ac.id

Abstract. The Multi Media Layer Coir Filter is a method of treating liquid waste by utilizing soil as the main medium which is shaped like a brick block and arranged alternately between zeolites. Multi Media Layer System made of acrylic (33x33x36) cm filled with brick blocks (12x7x4) cm mix of volcanic soil, charcoal, palm husk (7.5: 1: 1: 0.5), a permeable layer is used zeolite (1-3 mm). Samples came from the wells of the residents of housing Mendalo mas, Mendalo land in Jambi. The research was conducted with variations in flow rates of 10, 20, 30, 60 mL/minute, and two ways, namely aeration and non-aeration, the best efficiency of the aeration process was 10 ml/minute to reduce pH, odor, color, TDS, turbidity. Hardness, Nitrite, Nitrate, Mn, Fe, Cd, Sulfate, Cl, E. Coli at aeration 7,7, Odorless, 29,90%, 72,30%, 50,70%, 52,62%, 59, 70%, 57.30%, 66.40%, 83.22%, 21.70%, 37.53%, 28.88%, 100% and non-aerated 7,2, Odorless, 25.80% , 63.30%, 35.90%, 58.45%, 49.60%. , 45.49%, 46.25,% 78.59%, 18.75%, 37.21%, 28.18%, 100%. From all values, the reduction efficiency of all parameters is fulfilled for potable water.

I. Introduction

The volume of water in the human body is an average of 65% of the total body weight, and this volume varies greatly from person to person, even varies between parts of a person's body. Several organs in the human body that contain a lot of water, including the brain 74.5%, bone 22%, kidney 82.7%, muscle 75, 6%, and blood 83% [1][2]. Groundwater contains both inorganic and organic substances and is therefore a good place for microorganisms to live. The autotrophic microorganisms are the first inhabitants of water containing inorganic substances, The dead cells are an organic material that allows the life of heterotrophic microorganisms.

The Multi Media Layer System has several advantages, including being able to reduce the values of BOD, COD, TSS, color, and eliminating odors simultaneously, having a high ability to receive and absorb water flowed into the system, 1000 - 4000 Lm2d-1 while conventional soil 10 - 40 Lm-2d-1, very effective for removing organic and inorganic components from domestic wastewater. This is

[3] Deliza@uinjambi.ac.id.

Content from this work may be used under the terms of the Creative Commons Attribution 3.0 licence. Any further distribution of this work must maintain attribution to the author(s) and the title of the work, journal citation and DOI.
Published under licence by IOP Publishing Ltd

because the soil used is black andisol soil which contains aerobic and anaerobic microbes that can degrade dissolved organic compounds in water [3].

Mendalo mas is one of the locations of several housing estates in Mendalo Darat. The Mendalo mas area used to be a swamp, where the water was yellowish to reddish, tended to be cloudy and oily, acidic, when exposed to the skin it felt gummy and had a slight odor in the water bodies. the local community generally uses water from PDAM, but the provision and range of services from the PDAM is limited, to meet the water needs of the population using well water and in general, the well water that the community uses is not suitable, not under clean water standards, therefore it is necessary to treat it. well water before it is used for daily needs.

This research was conducted to see the ability of the combination of Multi Media Layer System - Coir Filters for oil palm bunches using volcanic soil material from the foot of Mount Kerinci, zeolite, palm shell charcoal (Kumpe, Jambi), iron powder, and coir of palm bunches (Kumpe, Jambi) in water purification. turbid and yellow wells to be fit for drinking by utilizing natural potential. Under the health quality standards listed in the Regulation of the Minister of Health of the Republic of Indonesia Number 492 / MENKES / PER / IV / 2010 [4].

2. Materials and Methods

2.1 The Material of the Multi Media Layer

The soil mixture layer of Multi Media Layer system consists of material, volcanic soil (from Kerinci), charcoal, palm husk (7.5: 1: 1: 0.5), permeable layer is used zeolite (1-3 mm).

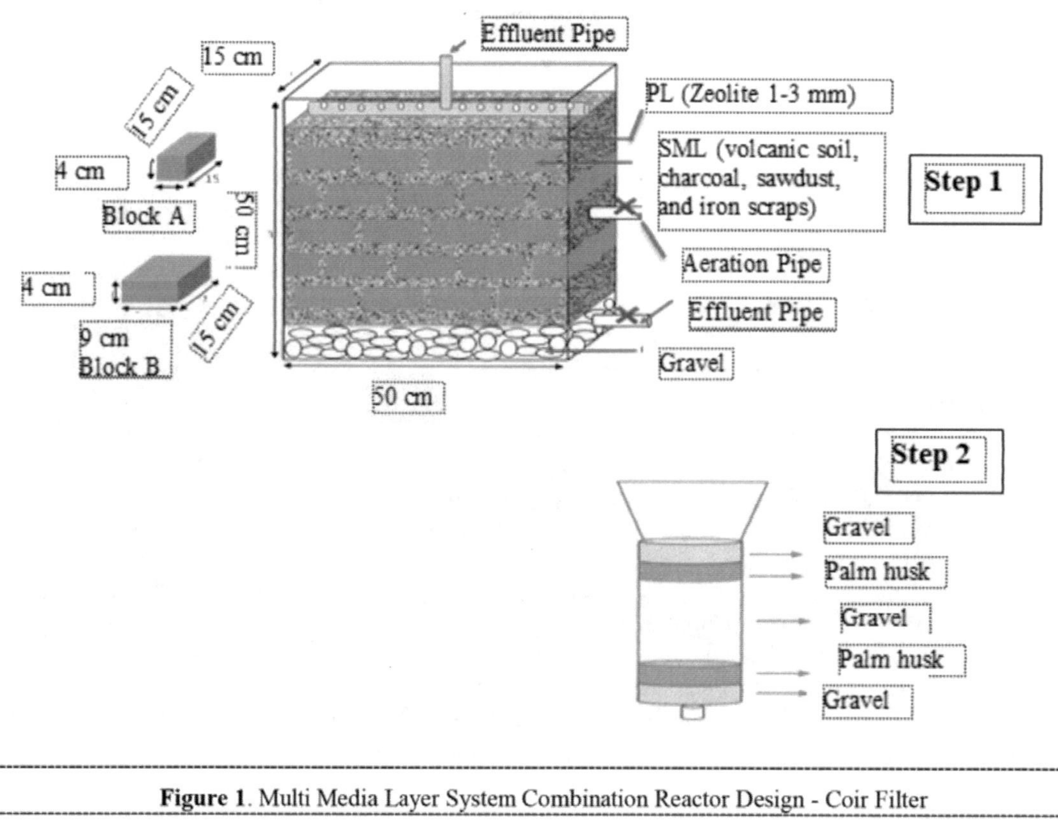

Figure 1. Multi Media Layer System Combination Reactor Design - Coir Filter

2.2 Treatment of the Sample

The well water that has been taken is done first by analyzing the water quality with physical parameters such as odor, turbidity, and several chemical parameters such as pH, Nitrite, Nitrate, metal identification, BOD, COD in the well water. To predict the capability of this filter media system, well water analysis is carried out before and after the aeration and non-aeration processes by varying the flow speed.

2.3 Analysis Method

All analysis of pH parameters, turbidity, nitrite, nitrate, Mn, Fe, Cd, TDS, Odor, Color, and E. Coli, according to the Indonesian National Standard (SNI) Determination of pH (SNI 06-6989.11-2004) [5], Determination Turbidity (SNI 06-6989.25-2005))[6]Determination of Nitrite Content (SNI-06-6989.9-2004) [7], Determination of Nitrate (SNI 06-2480-1991) [8], Determination of Manganese Metal (SNI 6989.5- 2009) [9]. Determination of Fe Metal (SNI 69-894-2009) [9] Determination of TDS, Determination of Odor, Determination of Color Content (SNI 69-8980-2011) [10],Determination of Cd Metal (SNI 2345-5-2011) [11] and E. Coli Determination [10].

3. Result and Discussion

Results of Well Water Analysis after Aeration and Non-Aeration Treatment in a multi layer system *Turbidity*. The results of turbidity analysis using aeration and non-aeration methods at flow rates of 10, 20, 30, 60 mL/minute still do not meet the drinking water quality standards but after an experiment by reducing the flow rate from 2, 4, 6, 8 mL /minute the turbidity value is already at the threshold value of Permenkes RI No. 492 / Menkes / Per / IV / 2010 [12].

The results of the analysis of the turbidity of well water after passing through the system in the aeration and non-aeration processes decreased in each flow rate variation (10, 20, 30, 60) mL / minute from the initial turbidity value of 27.455 NTU. However, the value for decreasing the turbidity is still above the standard value for drinking water quality. Then try again to drain the well water into the tool by reducing the flow rate to (2, 4, 6, 8) mL/minute in aeration and non aeration processes. The results were obtained that were already below the standard value of drinking water quality standards RI Regulation No. 492 / Menkes / Per / 2010.

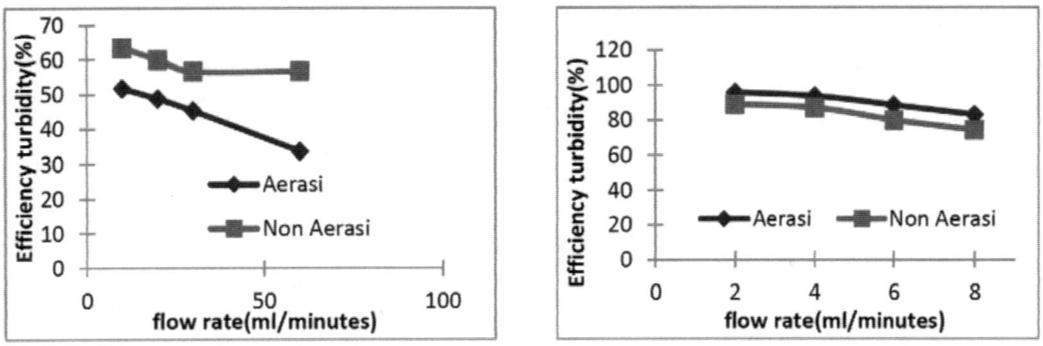

a	B

Figure 2. Graph analysis of turbidity in aeration and non-aeration, (a) Efficiency of reducing turbidity levels in the Multi Media Layer System method with aeration and non-aeration processes at flow rates (10, 20, 30, 60) mL / minute (b) Efficiency of reducing turbidity levels in the Multi Media Layer System-Coir Filter method with the process aerated and non-aerated at flow rates (2, 4, 6, 8) mL/minute.

IOP Conf. Series: Materials Science and Engineering **990** (2020) 012011 doi:10.1088/1757-899X/990/1/012011

Analysis Power of Hydrogen (pH)

The Power of Hydrogen (pH) is the concentration of hydrogen ion (H +) in a liquid. pH is also an important chemical parameter in water analysis. A water pH lower than 6.5 increases the corrosivity of metal objects, creates a bad taste, and can cause some chemicals to become toxic which can be detrimental to health [5].

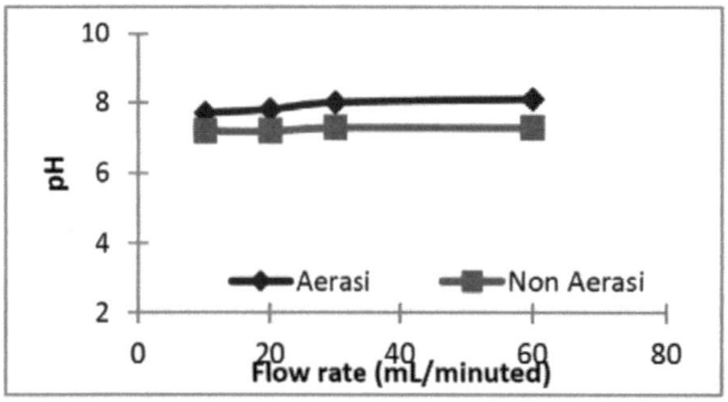

Figure 3. Graphic of pH values before and after treatment with the Multi Media Layer System-Coir Filter in aeration and non aeration processes.

Based on the pH value of well water (Figure 3) before passing it to the Multi Media Layer System-Coir Filter system is 6.49. This pH value is below the standard value of drinking water quality Regulation No. RI No. 492 / Menkes / Per / IV / 2010, namely 6.5-8.5. This means that the pH in the well water does not meet drinking water quality standards. From the results of measuring the pH of well water that has been passed to the Multi Media Layer System-Coir Filter system with aeration and non-aeration processes, the pH value increased from 6.49 to 7.2 - 7.7. This shows that the hydroxyl ion concentration (OH-) is higher than the hydrogen ion concentration. This is due to the increase in OH- ion from the denitrification process of Nitrobacter bacteria [4].

Color Analysis

The color of the water can be caused by the presence of organic and inorganic materials, due to the presence of plankton, humus, and metal ions (such as iron and manganese), as well as other materials. The presence of iron oxide causes the water to be reddish in color, the presence of manganese oxide causes the water to be brownish or black in color.

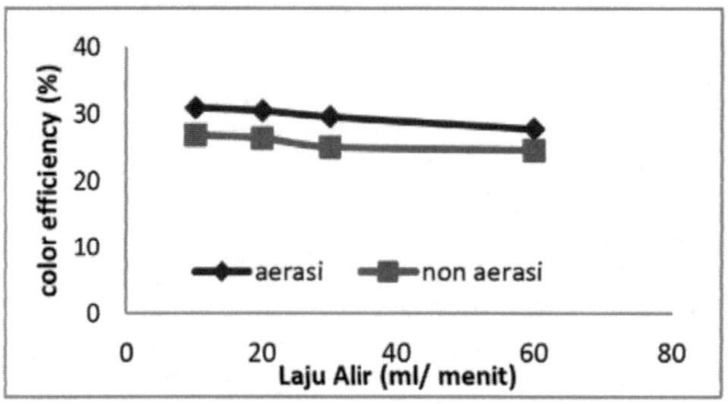

Figure 4. Graphic color efficiency

Total Dissolved Solids (TDS)

TDS consists of dissolved solids and suspended solids. The amount of dissolved solids consists of organic and inorganic compounds that dissolve in water, minerals, and salts. The effect of the total solid in the water consumed by the public will give an unpleasant taste to the tongue, nausea is mainly due to sodium sulfate and magnesium sulfate, and cardiac disease and toxemia occur in pregnant women and cause a decrease in the quality of health. The results of the TDS analysis can be seen in the figure 5.

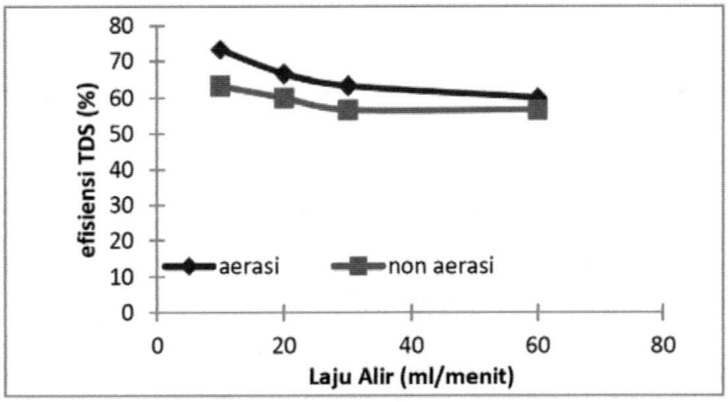

Figure 5. Grafic TDS Efficiency

Hardness

Hardness is a property of water caused by the presence of valence metal ions (cations), for example Mg2 +, Ca2 +, Fe + and Mn +. Total hardness is the hardness caused by the presence of Ca2 + and Mg2 + ions together

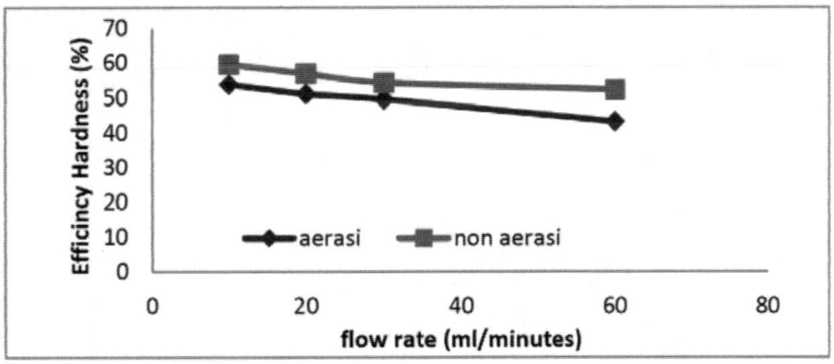

Figure 6. Hardness analysis efficiency graph

Nitrite Analysis

Nitrite This is a form between the oxidation of ammonia to nitrate or reduction of nitrate to ammonia Nitrite can enter the water through industrial wastewater. Nitrates can bind iron from hemoglobin which forms methemoglobinemia in the body.

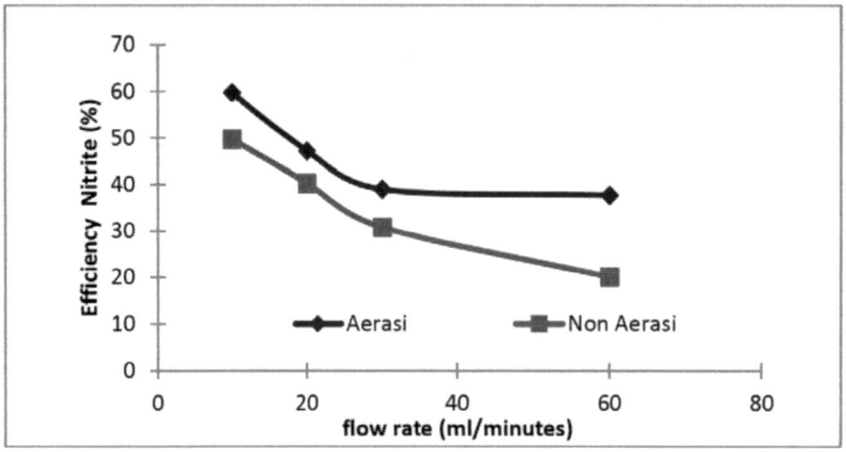

Figure 7. Graph of Nitrite reduction efficiency in aeration and nonaeration processes

The decrease in nitrite concentration that was detected at all variations in the flow rate indicated that the nitrification process took place under aerobic conditions. In that condition, nitrite is oxidized by Nitrosomonas bacteria to nitrate, and then nitrite in anaerobic condition can be reduced to nitrogen by facultative anaerobic bacteria such as Denitrobacilus and Nitosococcus.

The nature of nitrite which is unstable in the environment, will easily turn into nitrate in treated well water, and it is also possible that the total ammonia that is broken down into nitrite in the nitrification process is small, so that the resulting nitrite is low.

Nitrate Analysis

Nitrates are usually present in surface water in small concentrations, and may reach high concentrations in groundwater. The presence of NO3 in water is closely related to the Nitrogen cycle, in nature. In this cycle it can be seen that Nitrate can occur both from the N2 atmosphere and from the fertilizers used and from the oxidation of NO2- by bacteria from the Nitrobacter group.

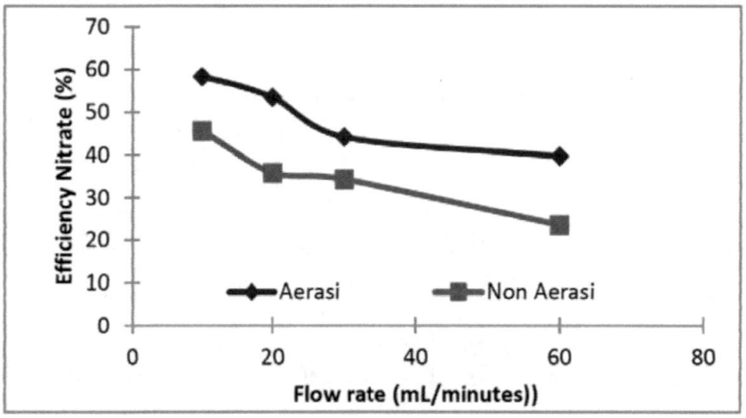

Figure 8. Graph of Nitrate reduction efficiency in aeration and non-aeration processes

IOP Conf. Series: Materials Science and Engineering **990** (2020) 012011 doi:10.1088/1757-899X/990/1/012011

In Figure 8, it is found that the Nitrate levels before and after being flowed to the of Multi Media Layer System - Coir Filters device are already at the quality standard value of drinking water in the Republic of Indonesia Regulation No. 492 / Menkes / Per / IV / 2010, namely 50 mg / L.

Mn Analysis

Manganese is one of the most abundant metals on the earth's surface, which is about 0.1% of the earth's crust. Manganese naturally occurs in surface and groundwater, but human activities also contribute a lot to contamination of manganese in water.10 Manganese can bind to nitrates, sulfates and chlorides and dissolves in water. MnO2 precipitates will stain white materials / objects. The presence of this element can cause odor and taste in drinks.

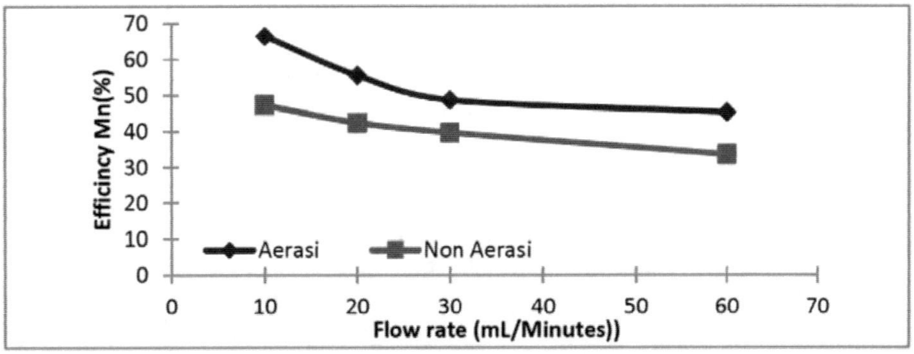

Figure 9. Efficiency graph of reducing Mn levels in aeration and non aeration processes

The value of the Mn content obtained also has met the standard quality value for drinking water standards RI Permenkes No.492 / Menkes / Per / IV / 2010, namely 0.3 mg / L. When compared to the level of efficiency of manganese reduction in aeration and non-aeration processes, the best is in the aeration process.

Fe Analysis

The presence of iron in water can cause water to become colored, smelly ,and tasteless. As many as 3 respondents (15%) stated that the well water smelled fishy and had a cloudy red color.

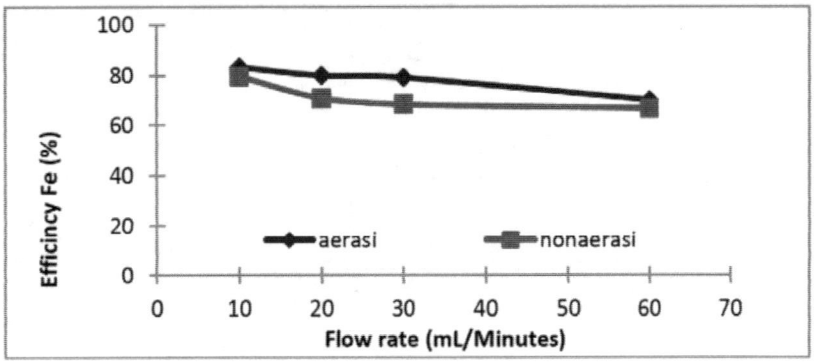

Figure 10. Graph of aerated and non aerated Fe Efficiency Analysis

Sulfate Analysis

Problems caused by the presence of sulfate in water are odor and corrosion problems in the piping resulting from the reduction of sulfate to hydrogen sulfide under anaerobic conditions. The laxative effect on sulfates occurs at concentrations of 600-1000 mg /L, if Mg + and Na + are cations that combine with SO_4, which will cause nausea and want to vomit.

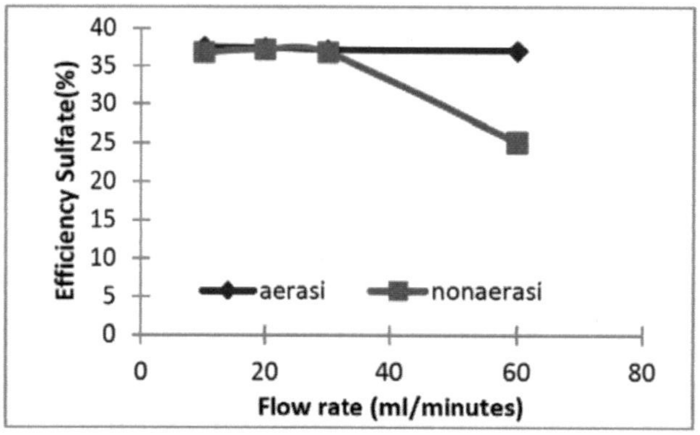

Figure 11. Graph of aerated and nonaerated Sulfate Efficiency Analysis

Cd Analysis

The content of cadmium (Cd) in clean water according to Permenkes RI No. 492 / Menkes / Per / IV / 2010, the maximum Cd level allowed in clean water is 0.003 mg / L. When compared with the Permenkes standard, the highest percentage of samples do not meet the requirements for clean water

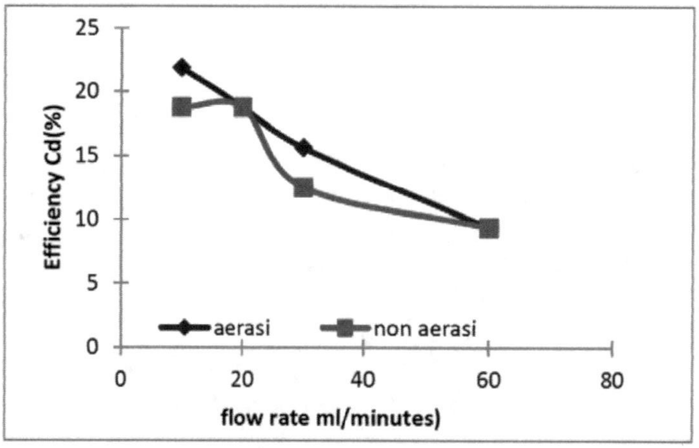

Figure 12. Graph of analysis of Cd efficiency in aerated and non aerated

Cl Analysis

Measurement of chloride content in the well water sample in Mendalo Jambi, this value is still below the threshold of the Ministry of Health's Quality Book Standard, which is 250 mg / L, and it can be seen that the of Multi Media Layer System - Coir Filters tool can reduce the Cl levels in well water.

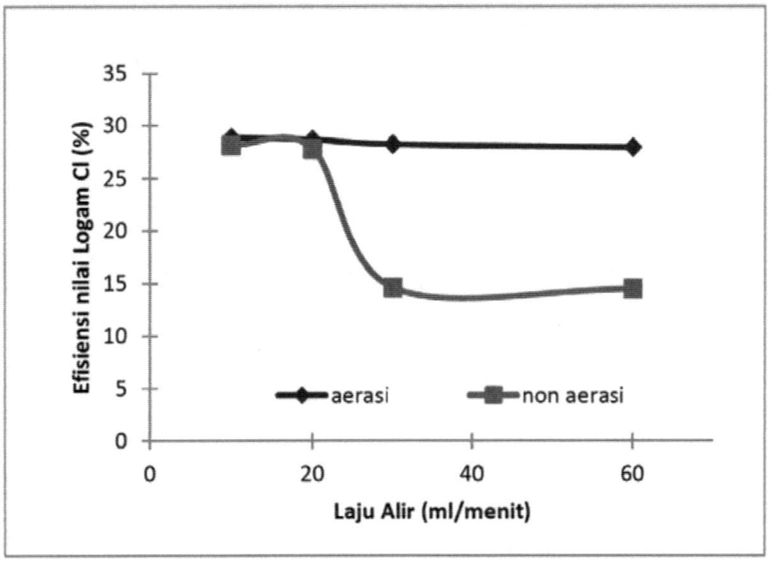

Figure 13. Graph of analysis of Cl efficiency in aerated and non aerated

E.Coli Analysis

These bacteria are active which utilize organic compounds as a growing medium. E. Coli grows at a temperature of 10-40 °C, with an optimum temperature of 37 °C and a pH of 4.0-9.0[13]. Size 2.0-6.0 µm in length and 1.1-1.5 µm in width, single arranged, paired with peritic flagella. The results of the analysis of E. coli bacteria [14] in well water can be seen in Figure 14 below

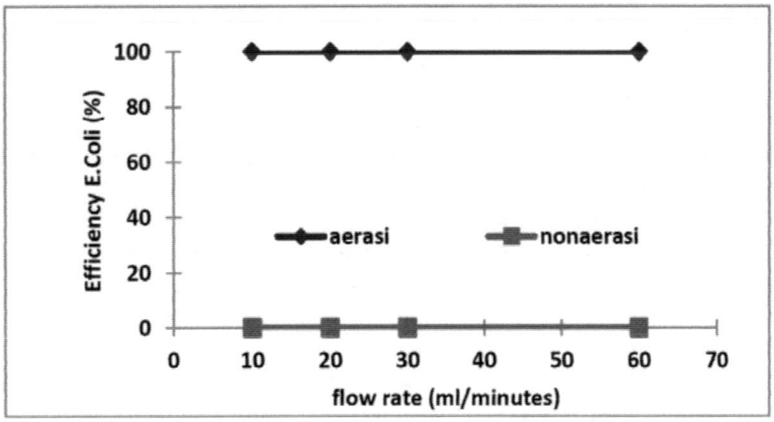

Figure 14.Graph analysis of the efficiency of E. Coli bacteria in aerated and non-aerated

The results of the analysis of the MPN (Most Probability Number) E. Coli before processing using the of Multi Media Layer System - Coir Filters have an average of 3 cells / 100 mL, after passing through the system in non-aeration processes with various flow rates (10, 20, 30, 60) mL / minute also obtained the same results as the initial value before flow into the appliance, namely 3 cells / 100 mL, did not experience a decrease in the value of E. Coli before being flowed into the appliance. This is due to the existence of various sizes of E. coli trapped in the tool, that in the river water there are many E. coli sizes that are bigger than well water E. coli, while the E. coli value of well water in the aeration process in all variations The flow rate was obtained by E. Coli bacteria of 0 cells / 100 mL. It has been a long time since the quality standard value of Permenkes No.492 / Menkes / Per / IV / 2010.

4. Conclusions

Based on drinking water quality standards according to Permenkes RI No. 492 / Menkes / Per / IV / 2010, the well water of a teak Padang resident before passing through the system, the turbidity parameter has not met the standard for drinking water suitable for consumption. The efficiency of the tool mixed with rice husks in well water treatment, for the aeration process at various parameters, the value of the reduction efficiency of all parameters is satisfactory for potable water.

Acknowledgment

The Monetary fund of this activity from Universitas Islam Negeri Sulthan Thaha Saifuddin Jambi was gratefully acknowledged

References

[1] J. Kesehatan and L. Kemenkes, 2011. 49–53.
[2] E. M. Widen *et al.*, "Validity of Bioelectrical Impedance Analysis for Measuring Changes in Body Water and Percent Fat after Bariatric Surgery," *Obes. Surg.*, vol. 24, no. 6, pp. 847–854, 2014, doi: 10.1007/s11695-014-1182-5.
[3] G. Yidong, C. Xin, Z. Shuai, and L. Ancheng, "Performance of multi-soil-layering system (MSL) treating leachate from rural unsanitary landfills," *Sci. Total Environ.*, vol. 420, pp. 183–190, 2012, doi: 10.1016/j.scitotenv.2011.12.057.
[4] Menteri Kesehatan RI No 492, "peraturan menteri kesehatan RI," pp. 1–9, 2010.
[5] S. N. Indonesia and B. S. Nasional, "Air dan air limbah – Bagian 11 : Cara uji derajat keasaman (pH) dengan menggunakan alat pH meter," 2004.
[6] S. N. Indonesia and B. S. Nasional, "Air dan air limbah – Bagian 25 : Cara uji kekeruhan dengan nefelometer," 2005.
[7] S. N. Indonesia and B. S. Nasional, "Air dan air limbah – Bagian 9: Cara uji nitrit (NO 2_ N) secara spektrofotometri," no. 2, 2004.
[8] "Pengujian Kadar Nitrat Dalam Air," p. 2480, 1991.
[9] S. N. Indonesia and B. S. Nasional, "Air dan air limbah – Bagian 18 : Cara uji nikel (Ni) secara Spektrofotometri Serapan Atom (SSA) – nyala," 2009.
[10] "Permenkes_492_tgl_19-April_2010_persyaratan_kualitas_air_minum.pdf." .
[11] S. N. Indonesia and B. S. Nasional, "Cara uji kimia – Bagian 5 : Penentuan kadar logam berat Timbal (Pb) dan Kadmium (Cd) pada produk perikanan," no. Cd, 2011.
[12] Peraturan Menteri Kesehatan RI, "PMK-RI No. 492/MENKES/PER/IV/2010 Tentang Persyaratan Kualitas Air Minum." 2010.
[13] A. K. Camper, G. A. McFeters, W. G. Characklis, and W. L. Jones, "Growth kinetics of coliform bacteria under conditions relevant to drinking water distribution systems," *Appl. Environ. Microbiol.*, vol. 57, no. 8, pp. 2233–2239, 1991, doi: 10.1128/aem.57.8.2233-2239.1991.
[14] T. M. Joyce, K. G. M. C. Guigan, J. E. T. Al, and A. P. P. L. E. N. M. Icrobiol, "Inactivation of Fecal Bacteria in Drinking Water by Solar Heating," vol. 62, no. 2, pp. 399–402, 1996.

IOP Conf. Series: Materials Science and Engineering **990** (2020) 012012 doi:10.1088/1757-899X/990/1/012012

Utilization Study of Carbonized Coal Briquette as Beef Rendang Cooking Fuel

Pasymi[1], E Sari[1] and E D Rahman[1]
[1]Chemical Engineering Department, Universitas Bung Hatta,19 Gajah Mada Street, Olo Nanggalo, Padang City (25143), Indonesia

E-mail : pasymi@bunghatta.ac.id

Abstract. Finding the cheaper alternative fuels and shortening the rendang cooking time are the potential efforts to reduce the production cost of rendang. This study aims to use low-rank cheap coal to cook beef rendang and to develop a more efficient rendang cooking method. First, the raw coals were carbonized, then the charcoals obtained were crushed and mixed with the filler and binder. After that, the charcoal mixture was molded into briquettes. Furthermore, the carbonized coal briquettes resulted were characterized and used as fuel to cook beef rendang in a semi-closed moveable-grate stove. To get a more efficient cooking time, a modified cooking method was developed. From the research has been obtained that there was almost no significant odor and smoke emissions during the cooking process except at the initial ignition period, which takes between 10-15 minutes. The modified cooking method took time 3 hours to get the dry beef rendang, 1 hour shorter than that of the normal cooking ones. Briquettes consumption for 1 kg of beef rendang cooking,using the modified cooking method, was about 0.75 kg, equivalent to USD 0.18 if 1 kg of briquette is priced USD 0.24. Finally, it can be summarized that carbonized coal briquette is a solid smokeless fuel that is safe and cheap to cook beef rendang.

1. Introduction

Besides being delicious, beef rendang is also very nutritious food. According to [1], the digested protein content in rendang beef increased by about 44% compared to fresh beef. Apart from containing high digested protein, rendang is also safe to consume. The saturated oil and fat content of beef rendang, which is dominated by long and medium-chain of fatty acids, are not easily degraded [1]. In addition, the presence of spices in the rendang mixture can inhibit the degradation of fatty acids [2]. The friendliness of fatty acids derived from coconut milk for human health has also been previously mentioned in [3].

The cooking process of rendang,ata cooking temperature ranging from 80-93 °C,takes time between 5-6 hours [2]. The rendang cooking process mechanism consists of water evaporating, converting coconut milk into oil, deactivating various enzymes present in the beef and coconut milk, and associating the spices in the rendang mixture.Dry beef rendang has a long shelf life. The lower the water content of beef rendang, the longer its shelf life will be and vice versa.

[1] pasymi@bunghatta.ac.id.

Content from this work may be used under the terms of the Creative Commons Attribution 3.0 licence. Any further distribution of this work must maintain attribution to the author(s) and the title of the work, journal citation and DOI.
Published under licence by IOP Publishing Ltd

The long cooking process of rendang (5-6 hours) cause fuel consumption to be high and increase the degradation potency of the rendang dough. This is one of the main problems of the rendang food business. This study aims to use low-rank cheap coal to cook beef rendang and to develop a moreefficientrendang cooking method. Various problems in using coal as fuel, such as initial ignition difficulty, smoke formation prevention, combustion control difficulty, and user comfort, as stated in [4], have been resolved in this study.

2. Methodology

The implementation of this research consisted of several stages, namely the briquette manufacturing stage, the briquette characterization stage, and the rendang cooking stage. The brief descriptions of each stage were given below.

2.1. Briquettes Manufacturing Procedure

The manufacturing flow diagram of the carbonized coal briquette used here was shown in figure 1. The low-rank coal from Muaro Jambi area, Jambi Province, Indonesia was carbonized by indirect heating in a multi-retort carbonizer for ± 6 hours. The charcoals obtained were then crushed, ground, and screened to get -8 mesh coal particles.The particles that pass through the 8 mesh sieve were fed to the mixing vessel to make the briquette dough while the rest was returned to the grinding process.

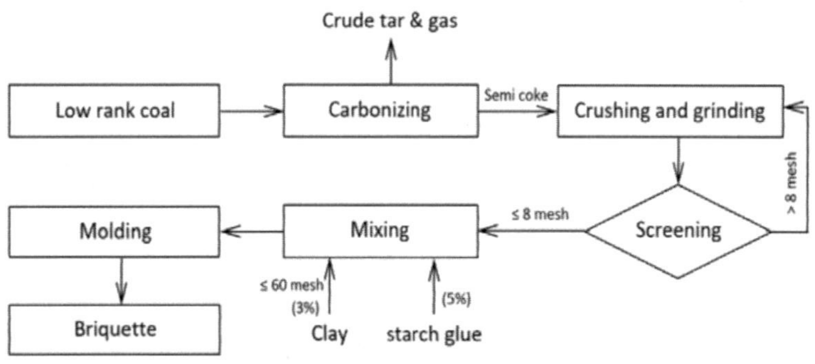

Figure 1. Flow diagram of the carbonized coal briquettes making process

The composition of the briquette mixture consisted of charcoal 92%, filler 3%, and binder 5%. The use of filleris intended to absorb gas emissions and to reduce the fly ash phenomenon during the cooking process. Meanwhile, the use of binder is intended to provide strength to the briquettes so that they are not easily destroyed during transportation, storage, and use[5]. The last step of the briquette making process was the briquettes molding, which was conducted manually.

2.2.The Briquette Characterizations

2.2.1. Physical Characterizations. One of the briquette physical characteristics determined here was density. The briquette density will correlate to the transportation and storage costs; the smaller the briquette density, the greater the transportation and storage volume required, resulting in higher handling costs. In addition, briquettes with small density tend to have low energy content. Conversely, if the briquette density is too high, the combustion performance of the briquette will be low due to the low porosity of the briquette. The quality standard of the briquette density is between 1.0 - 1.5 gr/ml [6].

The briquette density determination referred to ASTM D2395-17 [7]. The procedures were as follows: (a) weighing the mass of a briquette, (b) putting a briquette into a fully filled water vessel, (c) determining the volume of water that is removed from the vessel and record it as the volume of a briquette, (d) calculating the briquette density through equation (1); where ρ is briquette density, *m* is briquette mass and *v* is briquette volume. The briquette density was calculated based on an air-dried basis (adb).

$$\rho = \frac{m}{v} \tag{1}$$

Another physical property of the briquettes determined in this study was the shatter index.The shatter index indicated the briquette strength;whether the briquettes could stand or not during transportation, storage, and usecan be assessed from the shatter index value. For domesticfuel briquettes, the minimum shatter index required is 0.9 [8].

The determination of the shatter index referred to ASTM D440-86 [9]. The procedures for determining the shatter index were as follows; (a) weighing the initial weight of a briquette, (b) dropping the briquette by gravity from a height of 2 meters onto the metal floor 3 times, (c) weighing the residual weight of the briquette that has been resisted in an 8 mesh sieve, and (d) calculating the shatter index of the briquette through the equation (2).Variable I_s is a shatter index, m_{BI} is an initial mass of briquette, and m_{BR} is a residual mass of briquette [10].

$$I_s = \frac{m_{BR}}{m_{BI}} \tag{2}$$

2.2.2.Proximate Analysis. The parameters determined in the proximate analysis of briquette were water, volatile matter, fixed carbon, and ash contents. Proximate analysis was carried out on an air-dried basis (adb). The determination of the briquettes heating value,using bomb calorimeter, was also conducted in this step.

2.2.3. Initial Ignition Time. The initial ignition time of the briquettes is the length of time it takes from the first time the briquettes are ignited until the fixed carbons of the briquettes wereburnt stably. The unit of time of thebriquettes ignition was expressed in minutes. There were two kinds of ignition materials used in this research, namely kerosene and dry wood branches.

The procedures of using kerosene as an ignition material were as follows: (1) soaking some briquettes into the kerosene liquid for ± 5 minutes, (2) placing the soaked briquettes in the briquette pile, (3) igniting the briquettes with a match, (4) waiting for the fixed carbon of the briquette was burnt stably, (5) recording the time required for initial ignition. If the briquette hasn't been burned yet until the ignition material runs out, inject the kerosene until the briquettes were burnt stably.

Meanwhile, for the dry wood branches ignition material, the procedureswere as follows: (1) chopping the dry wood branches up to 5-10 cm in sizes, (2) puttingthe chopped of dry wood branches in a briquette pile, (3) igniting chopped dry wood branches, (4) waiting for the fixed carbon of the briquettes was burnt stably, and (5) recording the time required for initial ignition. If the briquette hasn't been burned yet until the ignition material runs out, add some chopped dry wood branches until the briquetteswere burnt stably.

2.2.4.The HottestLayer Position. There are two placements of the hottest layer in the briquette stove that were tested in this study, namely placement at the top and at the bottom of the briquettes pile. If the stove combustion chamber is first filled with briquettes and ignited later from the top, then the hottest layer position of the briquettes pile is at the top (figure 2.a). Conversely, if the stove combustion chamber is ignited first at the bottom and filled later with briquettes until it is full, then the position of the hottest layer of the briquettes pile is at the bottom (figure 2.b). The position of the hottest layer at the briquette pile will affect the smoke emission of a briquette stove. An illustration of the hottest layer positions in the briquette stove were shown in figure 2.

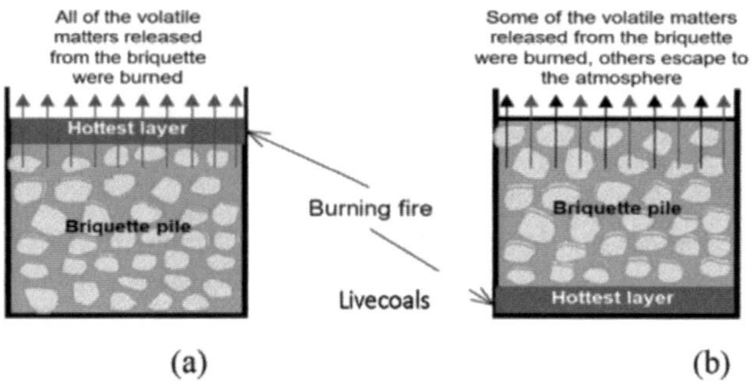

Figure 2. Illustration of the hottest layer position in the briquette pile

2.3. Cooking Test of Beef Rendang Using Briquettes

The type of stove used in the beef rendang cooking test was a semi-closed moveable grate stove, wherethe grate of the stove can go up and down, as shown in figure 3.According to [11], the recent problem of briquette stove design does not lie in the fuel combustion efficiency but lies in the heat transfer efficiency from the furnace to the cooking vessel.The moveable grate of the stove allows the hottest layer of the briquette pile always close to the cooking vessel so that the thermal efficiency of the stove would be better. The average thermal efficiency of this stove in the water boiling test was reported about ± 50% [12].

There were two cooking methods of beef rendang used in this study, namely: (1) a normal method, in which coconut milk and spices were cooked first until the dough releases oil (± 1 hour), then added the beef, and cooked until the dough dries, and (2) a modified method, where at first the beef and spices were sautéd with a little oil for ± 30 minutes, after that thick coconut milk was added and cooked until the dough dries. The time required to cook the beef rendang for both cooking methods was compared.

3. Results and Discussion

3.1. Physical Properties

The briquettes produced in this study was in the form of jengkol briquette with an equivalent diameter of about 41 mm. The appearance of the jengkol briquettes was shown in figure 4. The average density of the briquettes obtained was 1.17 gr/ml. This density value is within the density standard of fuel briquette, namely 1 – 1.5 gr/ml [6]. Meanwhile, the shatter index value of the briquettes obtained was about 96%. This value is above the minimum shatter index, namely 0.9 [8]. From these results, it could be seen that briquettes produced in this study have met the physical standard of the domestic fuel briquette.

3.2.Proximate Analysis

The proximate analysis results of the carbonized coal briquettes and quality standards of the fuel briquette[13], on the air-dried basis (adb), were shown in table 1. From the table, it could be seen that, in terms of moisture and calorific value, the values of proximate analysis of the coal briquettes produced have met the quality standards of fuel briquettes. But in terms of the volatility, the value was greater than the standard. Besides being caused by the incomplete carbonization process, the addition of tapioca flour was also the cause of the high content of volatile matters from the resulting briquettes. The addition of the tapioca flour used here was 5%, greater than that was recommended by Suganal, namely 3% [5].

Table 1. Proximate analysis of the carbonized coal briquette

No.	Parameter	Value		
		Experiment	SNI*	Suganal's
1.	Moisture, %	6.21	7.5	4.29
2.	Volatile matter, %	19.62	≤ 15	30.81
3.	Fixed carbon, %	57.05	n.d**	29.63
4.	Ash, %	17.12	n.d**	35.27
5.	Calorific value, kkal/kg	5.223	≥ 5000	4.412

* SNI = Standar Nasional Indonesia ** n.d = not be defined

3.3. Initial Ignition Time

The initial ignition time of the briquettes for the kerosene and wood branches was almost the same, ranging from 10-15 minutes. This time on average was slightly longer than the initial ignition time of coal briquettes reported by Suganal in [5], which was ± 10 minutes. This is because the volatile matters of the briquettes used here (19.6%) are lower than the volatile matters used by Suganal, which was 28.7%. The initial ignition time of the briquettes is directly proportional to the reactivity and the volatile matters content of the briquettes. The higher the content of the volatile matter of briquette, the higher its reactivity and the faster its initial ignition time [5].

Both types of ignition substances gave smoke emission at the initial ignition periode, where the smoke emission produced by the kerosene tended to be more than of the dry wood branches. The smoke formation occurred because of the volatile matters that come out from the briquettes were released into the environment without having time to burn in the stove combustion chamber. It was because the stove temperature at the initial use was still low. As time goes by, the temperature of the stove combustion chamber continued to rise, and the fixed carbon of the briquettes started to burn. As a result, the temperature of the stove combustion chamber was higher and the volatile matters of the briquettes would be burned in the combustion chamber before escaping to the environment.

3.4. Hottest Layer Position

From the two positions of the hottest layer that were varied in this study, it was found that the placement of the hottest layer at the top of the briquette pile was better in reducing the smoke emission than the placement of the hottest layer at the bottom. It was because when the hottest layer is at the top, the evolution of the volatile matters from the briquette will pass through and will be burned in the layer so that the smokes emission was not formed. Conversely, when the placement of the hottest layer is at the bottom of the briquette pile, the evolution of the volatile matters will escape to the environment without passing the hottest layer. As a result, a part of the volatile may be burned in the stove combustion chamber but others may be released to the environment as smoke.

Apart from being determined by the placement of the hottest layer position, smoke formation is also influenced by the stove design. The open stove system tends to produce more smoke than the stove with a closed system. This is because, in an open stove system, the evolution of volatile matters can leave the stove from many directions so that it does not always pass through the hottest layer of the stove. In addition, the hottest layer of an open stove system tends to have a lower temperature than a closed stove system, so it is not completely capable of burning the volatile matters that pass through it. The use of a semi-closed stove in this study gave a positive contribution to the absence of smoke.Even though the volatile matter content of the briquette used was slightly above the quality standard, but there is no smoke emission released from this combustion test.

IOP Conf. Series: Materials Science and Engineering **990** (2020) 012012 doi:10.1088/1757-899X/990/1/012012

3.5.Cooking Test Results of Beef Rendang Using Briquettes

The appearance of the dry beef rendang resulted from this study was shown in figure 5.From the two cooking rendang methods used here, it was obtained that the cooking time required for 2 kg of beef rendang, using the normal cooking method was ± 4 hours, while the cooking time required using the modified cooking method was about 3 hours. This happens because, in the normal cooking method, some time is sacrificed to produce oil from coconut milk after which the oil is associated with spices and beef[1]. Meanwhile, in the modified cooking method, the association of spices and meat is carried out using an external source of oil, so that this method requires less time.In terms of aroma and taste, the beef rendangs resulted from both of the cooking methods were similar, qualitatively. While, the protein and fatty acid contentsof the beef rendangs resulted from both of the cooking methods need further investigation.

Figure 3. Semi-closed Movcable grate stove

Figure 4. Typical jengkol briquette

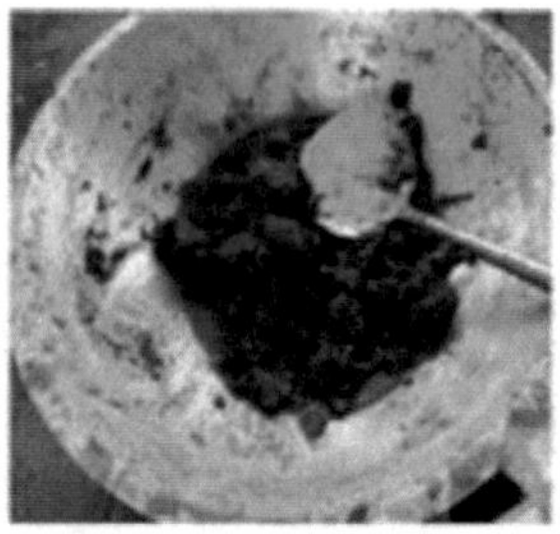

Figure 5. dry beef rendang

This study also revealed that the fuel consumption for cooking 2 kg of beef rendang by the modified cooking method was 1.5 kg of briquettes. While in the normal cooking method, the briquette consumed was 2.1 kg. If 1 kg of briquettes is priced at USD 0.24, then the fuel cost of 1 kg of beef rendang will be USD 0.18.

4. Conclusion

The cooking test of beef rendang using carbonized coal briquettes on a semi-closed moveable grate stove has been successfully carried out. Several results have been obtained, among them: (1) in general, the quality of the coal briquettes produced can meet the quality standards of fuel briquettes; both physically and chemically, (2) the initial ignition of coal briquettes, for both ignition materials, takes time between 10-15 minutes, (3) placing the hottest zone at the top of the briquette pile is more effective in reducing smoke emissions than placing it at the bottom, (4) using the briquettes as fuel in a semi-closed moveable grate stove did not emit smoke except during initial ignition, (5) cooking rendang using the modified cooking method was more efficient than that of the normal cooking method, (6) the quality of the beef rendangs resulted from both of the cooking methods needs further investigation, especially in terms of protein and fatty acid contents, (7) the length of time needed to cook 2 kg of beef rendang by the modified cooking method was 3 hours with a briquette consumption of ± 1.5 kg, and (8) if the price of the carbonized coal briquettes is USD 0.24 per kg, the fuel cost for cooking 1 kg of beef rendang is about USD 0.18. Finally, it could be concluded that cooking beef rendang using carbonized coal briquettes is safe and cheap.

IOP Conf. Series: Materials Science and Engineering **990** (2020) 012012 doi:10.1088/1757-899X/990/1/012012

References

[1] Rini, Azima F, Sayuti K and Novelina 2016 Agriculture and Agricultural Science Procedia**9** 335-341
[2] Azima F, Novelina and Rini2016 International Journal on Advanced science Engineering Information Technology**6** 465-468
[3] Lipoeto N I, Agus Z, Oenzil F, Wahlqvist M L and Wattanapenpaiboon N 2004 *Asia Pac J Clin Nutr***13** 377-384
[4] Pasymi, Sari E and RahmanE D 2012 National seminar proceeding ReSaTek **II** TK-E1-9
[5] Suganal 2009 Jurnal Teknologi Mineral dan Batubara**05**17 – 30
[6] Boasiako A and Acheampong B B 2016 Biomass and Bioenergy**85** 144 - 152
[7] ASTM D2395-17 2017*Standard Test Methods for Density and Specific Gravity of Wood and Wood-Based Materials;* ASTM International: West Conshohocken, PA, USA
[8] Borowski G, Stępniewski W and Oliveira K W 2017 *Int. Agrophys.* **31** 571 – 574
[9] ASTM D440-86 2002*Standard Test Method of Drop Shatter Test for Coal*; ASTM International: West Conshohocken, PA, USA
[10] Kpalo S Y, Zainuddin M F, Manaf L A andRoslan A M 2020 Production and characterization of hybrid briquettes from corncobs and oil palm trunk bark under a low pressure densification technique*Sustainability12* 2468
[11] Olumethodomi J O, Modestus O O & Lagouge K T 2019 part a: recovery, utilization, and environmental effects 1-19
[12] Aprizon, Ardian B, Pasymi and MartynisM2013 Modification of the coal briquette stove *Abstract of Undergraduate Research* Faculty of Industrial Technology Bung Hatta University **1**(4)
[13] SNI 4931 2010 Coal briquette: Indonesian National Standardization Body

IOP Conf. Series: Materials Science and Engineering **990** (2020) 012013 doi:10.1088/1757-899X/990/1/012013

Design of Biomass Briquette Stoves: Performance Based on Mixed of Durian Bark, Coconut Shell and Palm Shells as Materials of Bio Briquette

E Sari[1], Burmawi[1], U Khatab[2], ED Rahman[1], A PAnindi[1], E Andriyati[1], IAmri[1]
1 Faculty of Industrial Technology, Bung Hatta University, Padang, West Sumatra
2 Department of Civil Engineering Payakumbuh College of Technology, Payakumbuh, West Sumatra

E-mail: sariellyta@gmail.com

Abstract. Biomass is a renewable energy source derived from plants and is known as green energy. Biomass is a term used for various types of organic matter in solid form that can be used as fuel. Bio Briquette is a solid fuel made from a mixture of biomass. This fuel is an alternative material that was developed in a bulk with relatively short time and is relatively cheaper. The use of briquettes that must be included with the use of a stove or stove which type and size must be adjusted to the needs. To increase the heat efficiency of the existing briquette stoves, this research was carried out using two different types of stove walls, namely cement-filled stoves and glass wool-filled walls. The performance of the stove can be seen from the combustion of biomass briquette fuel against the briquette shape was molded. From the tests that have been carried out, the efficiency of stoves with cement walls is 29.86%, while for stoves with glass wool walls is 40%. Of the three forms of bio briquette (ellipsoidal, cylinder, perforated cylinder) the use of cylindrical bio briquettes is better because the flame on the briquette is longer when compared to other forms.

1. Introduction

Briquettes are a renewable alternative energy that has great potential in Indonesia. Consider an agricultural country where there are many raw materials for briquettes in large quantities.When compared to oil raw materials which are now classified as rare and non-renewable, briquettes have their own advantages. Meanwhile, bio briquette is a source of biomass energy that is environmentally friendly and also biodegradable.One of these is biobriquettes from biomass.Biomass is composed of cellulose, hemicellulose and lignin that is commonly found in plant parts [7]. Biomass is a source of energy for the future that will never run out, even the number will always increase, so it is very suitable as a source of fuel for households [1]. The biomass that used in this research is durian peel with mixed biomass coconut and palm shells. The addition of coconut shell and palm shell biomass improve the quality of briquettes durian peel [8]. The making of biobriquettes is carried out by the carbonization process, which is the process of making charcoal. The optimum temperature and time in the carbonization process for making biobriquettes are 400°Cand 2 hours [9].In the biobriquettes manufacture, carbonization is a very important process because it is the main process in the biobriquettes manufacture whichcan affect the biobriquettes quality [7].The forms of biomass briquettes used are ellipsoidal, perforated cylinders and cylinders [10]. From the three forms of biobriquette (ellipsoidal, cylinder, perforated cylinder) which has the best compressive strength, density and ignition rate is the cylinder shape.The used of briquettes must be accompanied by the use of a stove or stove with the type and size of the furnace that must be adjusted to the needs. It is just that briquette stoves that have been sold in the market have not attracted the public's interest due to their complicated use and also the flue gas produced from the briquette burning process can harm

[1] sariellyta@gmail.com.

Content from this work may be used under the terms of the Creative Commons Attribution 3.0 licence. Any further distribution of this work must maintain attribution to the author(s) and the title of the work, journal citation and DOI.
Published under licence by IOP Publishing Ltd

health and besides that it also damages kitchen utensils because it leaves black spots on cooking utensils. In previous research, the stove produced still had shortcomings.

One example is the research on the design of the bio briquettestove Sudip which has disadvantages such as baked ceramic stoves and burnt stoves easily cracked, and all that becomes a problem is the ash remover place that is not available so that it does not last long and cannot continue burning[5]. For the design of this briquette stove, it aims to get a high calorific value with maximum efficiency. Maximum in terms of minimum heat loss value, minimum exhaust gas loss, minimizing heat loss due to material.

2. Methods

The methodology used in this study is an experimental method, testing each variation of the amount of charcoal, ignition time, and cooking time. The resulting of bio briquette stoves are cement walls and *glasswool*. For *glasswool* stove specification details are ; the stove equipped with jack with overall height is 70 cm, wall stove height is 40 cm , diameter is 26 cm. The process of measuring data is carried out by determining the initial volume of water and the initial temperature of the water. Determining the amount of bio briquette with various shapes of cylinders, perforated cylinder, ellipsoidal and mixtures.

The tests carried out were the performance of the bio briquette stove, the efficiency of the stove, and the effect of the shape of the bio briquette on temperature and cooking time. Flowchart of the testing phase can be seen in Figure 1.

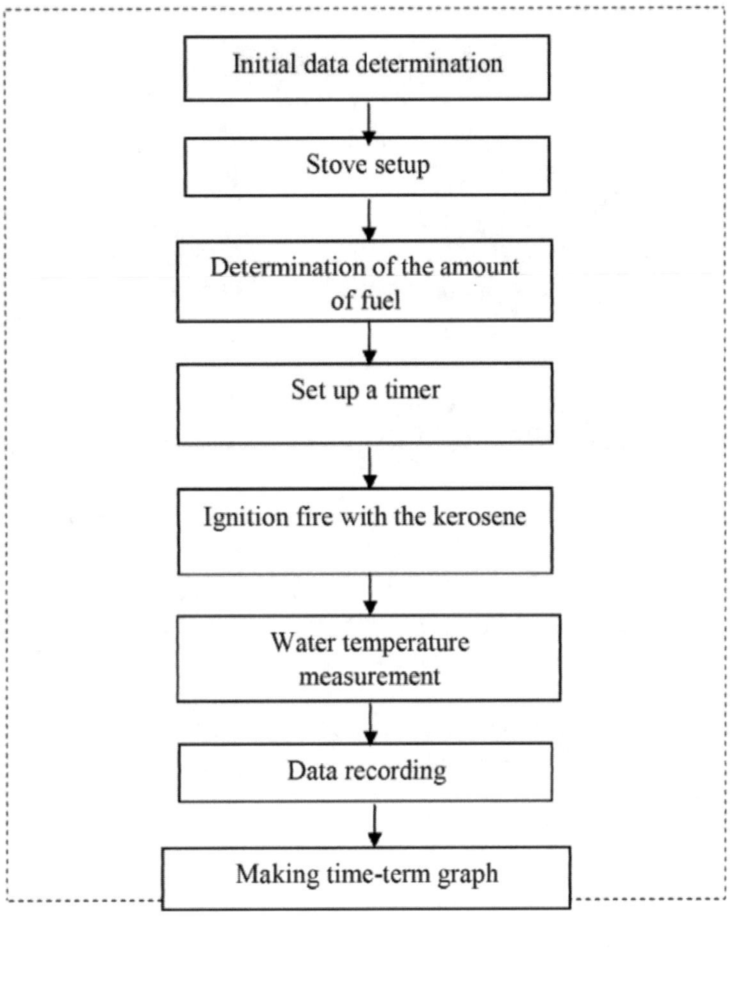

Figure 1. Flowchart of the testing phase.

3. Results and Discussion

The results of the biomass briquette stove design can be seen in Figure 2.

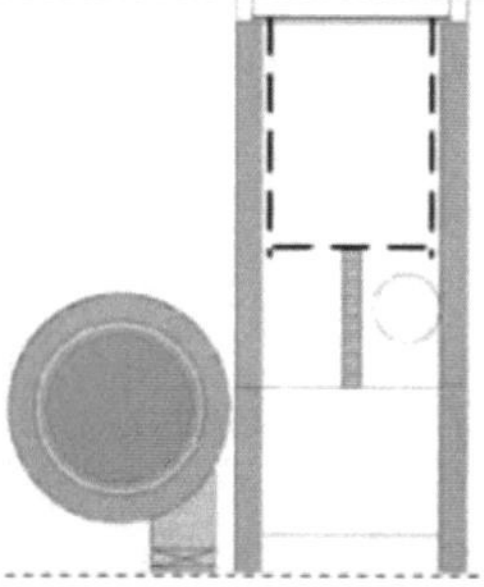

Figure 2. Bio Briquette stove design

The design results of the biomass briquette stove can be seen in Figure 3.

(a) (b)

Figure 3. Design of biomass briquette stove results ; (a) Stove top view, (b) Stove side view

From the results of the design that has been done, there are two variations of the stove wall, namely glass wool and cement, by varying the stove wall in order to see the inhibitor of heat loss from biomass briquettes so that the heat generated can be completely absorbed by water.

This biomass stove has a fan that is located on the side of the stove which is used to blow air from the bottom up so that the resulting fire does not smoke. Then on the stove there is a jack which functions to facilitate the rise and fall of the fire used. Furthermore, on the stove, there is a hollow biomass container or place where the ash after burning will fall and not accumulate in the container. And beside the stove wall there is an ash holder that can be opened so that it can be easier to clean. The performance of the bio briquette stove can be seen in Figure 4.

Figure 4. Bio Briquette stove performance

Figure 4. is the result of turning on the bio briquette stove. The resulting fire does not produce smoke when there is an air supply from the fan so that the combustion of bio briquettes is more complete than the absence of air supply.

According to another researcher,the combustion process that occurs often results in incomplete combustion so that the combustion that occurs produces carbon monoxide, so to obtain complete combustion, sufficient oxygen must be met [3]. The oxygen used can be pure oxygen or oxygen from the air.

The calculation data for the efficiency of the biomass briquette stove can be seen in Table 1

Table 1. Calculation data of the Efficiency of the Biomass Briquette Stove

Spesification	Data 1		Data 2		Data 2	
	Cement Wall Stove	Glass Wool Wall Stove	Cement Wall Stove	Glass Wool Wall Stove	Cement Wall Stove	Glass Wool Wall Stove
Initial briquette weight (g)	316	319	350	350	350	350
Weight of briquettes after burning (g)	209	188	210	209	206	209
Initial volume of water (ml)	500	500	500	500	500	500
Final volume of water (80°C)(ml)	445	466	461	455	477	464
Weight of remaining charcoal (g)	141	137	142	157	136	157.1
Weight of used charcoal (g)	68	51	68	52	70	51 , 9

From the experiments and calculations that have been carried out, the efficiency value of each bio briquette stove is obtained, namely, for the bio briquette stove with a variety of stove walls using for cement valued at 29.86%, while for bio briquette stoves with a variation of the stove walls, glass wool was valued at 40%. Compared to the variation of cement stove walls, the efficiency of glass wool stove wall variations is much higher. The thermal conductivity value of cement and glass wool has respective values of 1.73 and 0.038 W / m°C which means that the ability to conduct heat for cement is greater than that of glass wool, so that glass wool is more can reduce the heat loss [2].

In testing the ignition of the bio briquette stove using variations in the form of ellipsoidal bio briquettes, cylinders and perforated cylinders which can be seen in Figure 5.

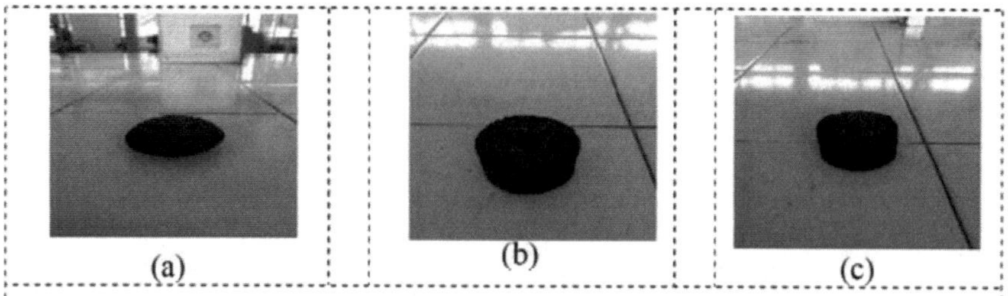

Figure 5. The shapes of bio briquettes ; (a) Ellipsoidal shape, (b) Perforated Cylinder shape,(c) Cylinder shape

This testing is carried out from boiling water until the briquettes run out, the results can be obtained. seen in Figure 6.

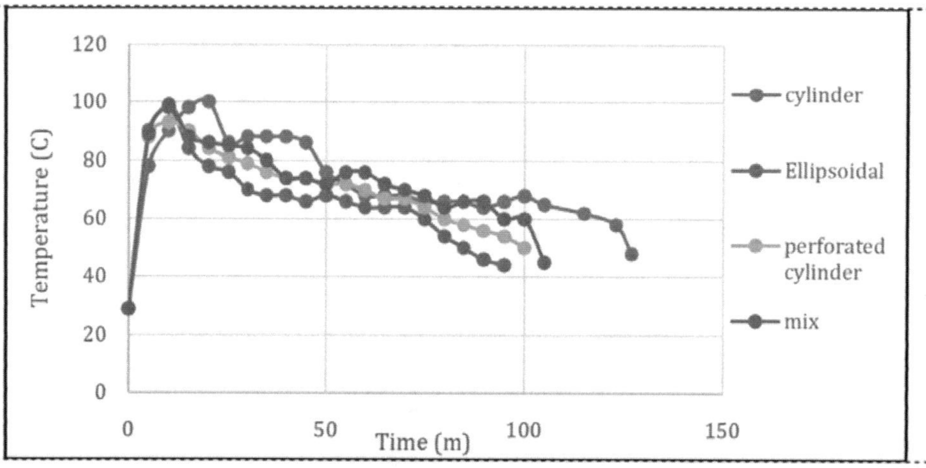

Figure 6. The effect of briquetes shape on cooking time

In Figure 6, it is the result of the trial results of the long ignition of the bio briquette stove using variations in the form of briquettes weighing 200 grams to boil 300 ml of water. From this experiment, the highest temperature ellipsoidal briquette in water was obtained at 98 °C, with 10 minutes cooking times. The length of time to turn on the stove in the ellipsoidal form until the charcoal runs out is 95 minutes with the result that the final volume of water is 79 ml, the final temperature is 44 °C.

For variations in the shape of cylindrical briquettes, the highest temperature that can be achieved is 100 °C at 20 minutes of turning on the stove. The charcoal runs out for 127.19 minutes with a final volume of water 69 ml and a final temperature 48 °C of water. For variations in the shape of the perforated cylinder briquettes, the results obtained are the highest temperature of 93 °C in the 10 minute of stove ignition. The ignition time for this hollow cylinder shape is 109.41 minutes. The final water temperature was 50 °C with a final volume of 49 ml. For the last trial, it was carried out with a mixture of the three variations of briquette shapes, namely ellipsoidal, cylinder and perforated cylinder. The ignition of this mixed variation briquette, the stove ignition time reached 124.12 minutes, the final temperature obtained was 45 °C with the final volume of water as much as 35 ml.

IOP Conf. Series: Materials Science and Engineering **990** (2020) 012013 doi:10.1088/1757-899X/990/1/012013

Compared to the results of the four experiments above, the shape of this cylindrical briquette takes longest to ignite. This is because perforated briquettes have a wider oxygen flow space, therefore combustion occurs more easily. The flow of oxygen in the hollow briquette occurs on the outside and inside of the briquette so that the briquette burns out faster. Whereas in cylinder briquettes, oxygen flow only occurs on the outer contents of the briquettes. Another study explaine the reduced air cavity in briquettes with a higher density will slow down the rate of combustion [4]. During the cooking process, a distance from the fire to the boiling pot must be kept. To maintain this distance, this is done by raising the shell where the briquettes are ignited by using a jack.

4. Conclusion

The results of the research that have been carried out provide the following conclusions:

- The resulting bio briquette stove design is a stove with glass wool-coated walls.
- In the stove, it is added with holes in the shell to remove ashes so that it does not accumulate, as well as an air supply from the fan.
- The bio briquette stove that has better efficiency between cement wall variations and glass wool coated walls is a stove with glass wool coated walls, which is 40%.
- Of the three forms of bio briquette (ellipsoidal, cylinder, perforated cylinder) the use of cylindrical bio briquette is better because the flame on the briquette is longer when compared to others.

Acknowledgments

We would like to thank:

- Directorate of Research and Community Service, Directorate General of Research and Development, Ministry of Research, Technology and Higher Education,
- Chancellor and Chair of LPPM Bung Hatta University for the opportunities and facilities provided by
- All parties involved in this research, lecturers, students and workshops from Bandung who helped carry out this research.

REFERENCES

[1] Basriyanta. 2007. *Harvesting Garbage*. Yogyakarta: Kanisius

[2] Buchori, Luqman. 2004. *Heat Transfer Part I*. Diponogoro University, UNDIP. Semarang

[3] Farah 2012 University of Indonesia, UI, Depok

[4] Nona Merry M.Mitan, Akmal H A, NurFathiah M N and Sian Meng Se 2015 Applied Mechanics and Materials **761** 494-498

[5] Pandey Sudip et al 2013 *Analysis and Test of Biomass Briquette and Stoves*. Nepal Academy of Science and Technology

[6] Pramudya, Sari E, dkk2013E-jurnalBung HattaUniversity UBH Padang

[7] Riseanggara, R.R 2008 *Optimization of Adhesive Levels in Waste Biomass Briquettes*. Faculty of Agricultural Technology, IPB. Bogor

[8] Sari E, Pasymi, Khatab U, Desmiarti R, Ariansyah R, Hariadi and Maisaroh S 2018IOP Conf. Series: Mat Sci and Eng**316** 012021 doi:10.1088/1757899X/316/1/012021

[9] Sari E, Burmawi, Khatab U, Rahman ED, Humaira A,Muflihah I, Amri I, Andini AP 2019International Journal of GEOMATE (submit)

[10] Sari E, Humaira A, Yusri F. 2019.*Effect of pressure and mold shape on the quality of the biobriquette from the durian peel biomassmixture.*Research report. Bung Hatta University, UBH. Padang

IOP Conf. Series: Materials Science and Engineering **990** (2020) 012014 doi:10.1088/1757-899X/990/1/012014

Characteristic of Tourim Solid Waste of Harau Valley, West Sumatra

R Aziz[1], Mahmuda[1], Y Ruslinda[1], Y Dewilda[1]
[1]Environmental Engineering Department, Faculty of Engineering, Universitas Andalas, Kampus Limau Manis Padang 25163, Indonesia

E-mail: rizkiaziz@eng.unand.ac.id

Abstract. Harau Valley Tourism Region is one of the three leading tourist attractions in Lima Puluh Kota Regency. The number of visitors to the Harau Valley Tourism Area has increased every year which will also have an impact on the increase in the amount of waste generated in this region. The Harau Valley tourist area does not yet have solid waste data. Basic data is needed to plan a waste management system in an area. Based on this, research is needed regarding the generation, composition and potential of waste recycling in the Harau Valley Tourism Area. The research method used is direct measurement in the field with reference to SNI 19-3964-1994. Based on the research that has been done, the generation unit is 0.158 kg/cap./day in weight units and 3.053 Lit/cap./day in volume units. The waste composition consisted of food waste (34.29%), paper (8.39%), plastic (16.43%), textiles (1.12%), rubber (0.07%), yard waste (17, 05%), wood (12.74%), glass (3.08%), non-ferrous metals (0.92%) and others (5.91%).Recyclingpotentialforfood and yardwetwaste (77.26%), plastic (70.23%), paper (56.21%), and glass (95.69%)Itisrecommendedtoapplycompostingmethodforcompostablewaste, whileforotherrecyclablewaste can applycollection, separation, clensing, and packingbeforesellingintonearbyshanties.

1. Introduction

Harau Valley is one of the leading tourist attractions in West Sumatra, located in Lima Puluh Kota Regency. Initially this attraction was part of the Harau Valley Nature Reserve, then designated as the Harau Valley Tourist Park based on Minister of Agriculture Decree No. 478 in 1979 with an area of 27.5 hectares. Based on a joint agreement with the Provincial Head of Region with the regencies and cities in West Sumatra in 2006, this area designated as a Tourism Development Area. Meanwhile, Regional Regulation No. 7 of 2012 concerning the Lima Puluh Kota Regional Spatial Plan [1], it was stated that the Harau Valley Tourism Area is one of the three leading tourist attractions in the Lima Puluh Kota Regency.

The number of tourists visiting the Harau Valley Tourism Area always increases every year. The number of tourists visiting Lima Puluh Kota Regency in 2017 was 303,990 people with an average daily visit of ± 630 people [2]. However, the high level of tourist visits and activities were notaccompanied by a good waste management plan from the local government. As a result, waste management in the Harau Valley tourist area is still not well practiced, this can be seen from the

[1] rizkiaziz@eng.unand.ac.id

Content from this work may be used under the terms of the Creative Commons Attribution 3.0 licence. Any further distribution of this work must maintain attribution to the author(s) and the title of the work, journal citation and DOI.
Published under licence by IOP Publishing Ltd

IOP Conf. Series: Materials Science and Engineering **990** (2020) 012014 doi:10.1088/1757-899X/990/1/012014

availability of inadequate facilities and infrastructure to accommodate the quantity of waste generated from this tourist area. In addition, the lack of public and visitor awareness of environmental cleanliness results in the large amount of litter dumped in the Harau Valley Area, Lima Puluh Kota District. This will certainly affect the beauty, public health and disrupt the quality of the surrounding soil and groundwater.

Tourism activities are growing rapidly in the world. The development of tourism contributes impacts on the surrounding environment, positively and negatively. Positive impacts such as the local employment and the income from tourism can build regional infrastructure. Negative impacts include waste problems, air pollution, water pollution and damage of natural environment on tourist destination. Some studies focus on solid waste management in the tourism area has took places on some countries includes Kashmir [3], Cyprus [4], Spain [5], Vietnam [6], Romania and Italy [7], and Indonesia.

Based on the survey results, the waste management system in the LembahHarau Tourism District of Lima Puluh Kota is still not good. This showed from the habits of the people who manage waste using the old paradigm method, namely collecting, transporting and disposing. Therefore, it is necessary to plan a waste management system in the LembahHarau Tourism District of Lima Puluh Kota in an environmentally sound manner. Before a waste management plan is designed the LembahHarau Tourism Area, Lima Puluh Kota Regency, it is necessary to carry out measurements, analysis of the generation, composition, and potential of waste recycling generated from the Harau Valley Tourism Area. With the implementation of an environmentally friendly waste management system in the Harau Valley Tourism Area, the amount of waste generation transported to the landfill will be reduced, and can create and maintain cleanliness and beauty conditions that will affect the sustainability of the Harau Valley tourist attraction.

2. Methodology

The research stages are as follows: 1) preparation, including studies and related literature to deepen and sharpen the basic theories relating to tourism waste management systems; 2) secondary data collection, secondary data is data obtained from agencies related to research, such as: Map of the area according to RTRW, existing technical and non-technical aspects d number of visitors; 3) primary data collection, data was taken through direct measurements in the field such as: generation, composition and potential of waste recycling;Sampling for primary data is carried out for 8 consecutive days referring to SNI 19-3964-1994 [16]. Sampling was carried out in the Harau Valley Tourism Area. The location and number of sampling points can be seen in table 1.

Solid Waste Generation

Waste generation can be expressed in units of weight (kg / person / day, kg / m2 / day) and volume units (liters / person / day, liter / m2 / day). To get a more representative calculation, sampling was carried out for 8 consecutive days at 24 sampling points spread in the Harau Valley Tourism Area.

$$\text{Generation formula:} = \frac{\text{weight or volume of waste (kg)}}{\text{area or number of waste producer (m}^2\text{ or person)}} \quad (1)$$

Solid Waste Composition

The first thing to do to calculate the composition of waste is to separate waste based on its components, namely dry waste and wet waste. Percentage of composition is the weight of each waste component divided by the total weight of the overall waste. Furthermore, the data can be used to calculate the potential for waste recycling based on the source and based on the components forming the waste.

Table 1. Location and number of sampling

No	Facility	Number of facility	Number of Sample
1.	Restaurant	4	1
2.	Shop		
	1. Food	130	13
	2. Accessories	10	1
	3. Clothes	15	2
3.	Waterfall area	3	1
4.	*Camping Area*	2	1
3.	Parking area	2	1
4.	Garden	3	1
5.	Mosque	3	1
6.	Toilet	3	1
7.	*Homestay*	3	1
Total		178	24

$$\text{Composition formula (\% weight): } = \frac{\text{weight of waste component (kg)}}{\text{total weight of waste (kg)}} \; \boldsymbol{x\ 100\%} \quad (2)$$

Solid Waste Recycling Potency
Recycling potency of solid waste determines by further sorting of waste which classify its into recyclable and non recyclable wastes. The percentage of recyclable waste defines the recycling potency

$$\text{Waste recycling potential formula (\%): } = \frac{\text{weight of the recyclable waste}}{\text{total weight of waste}} \text{ x } 100\ \% \quad (3)$$

3. Results and Discussions

3.1. Waste generation
From this study, the average solid waste generation in the Harau Valley Tourism Area is 0.158 kg/cap./d for weight units and 3.053 Lit/cap./d in volume. The details of the generation of waste from each source of waste can be seen in Table 2.

By obtaining waste generation based on area, the calculation of waste generation unit is based on visitors by dividing waste generation that has been obtained by the average number of visitors per day. This was done because:
1. Visitors who are producers of waste in the tourist area produce not only one source of facilities, but can also produce garbage at two or more facility sources, so that the waste generation unit data per source cannot represent all the waste generated by visitors.
2. Based on the analysis of the number of waste generation results of the multiplication unit of waste generation with the number of visitors and or the area of the waste producing area, the generation of waste generated by the number of visitors, is smaller than by area.
The average solid waste generation unit in the Harau Valley Tourism Area is 3.053 Lit/cap./d or 0.158 kg/cap./d.

Table 2. Solid waste generation

Facility	Area (m^2)	Generation per unit ($kg/m^2/d$)	Generation (kg/d)
Shops	1,000	0.086	86.341
Garden	1,000	0.003	2.906

Mosque	192	0.025	4.849
Toilet	60	0.029	1.759
Waterfall area	520	0.046	23.778
Restaurant	96	0.086	8.217
Parking area	1,000	0.001	1.367
Homestay	108	0.009	1.010
Camping Area	1,500	0.001	0.623
Total	5,476	0.286	130.850

3.2. Waste composition
The composition of waste in the Harau Valley Tourism Area is different which consists of wet waste, dry waste and others. The largest composition of waste generated from the Harau Valley Tourism Area is leftover food waste with a percentage of 34.29%. This is because this area is dominated by food traders. The composition of waste in the Harau Valley Tourism Area can be seen in table 3.

Table 3. Composition of solid waste

Components	weight (kg/day)	%
Food waste	11.653	34.29
Paper	2.851	8.39
Plastic	5.583	16.43
Textile	0.382	1.12
Rubber	0.024	0.07
Yard waste	5.796	17.05
Wood	4.329	12.74
Glass	1.046	3.08
Non Ferrous metal	0.314	0.92
Other	2.009	5.91
Total	33.977	100.00

3.3. Recycling potential
Calculating the potential for recycling is useful for determining what processing will be carried out at the Solid waste treatment station. The potential of recycling waste in the Harau Valley Tourism Area is quite good, for the highest potential for recycling compostable waste with a percentage of 41.31% of the total garbage available. Details of recycling potential can be seen in Table 4.

Table 4. Recycling potency

Composition	(kg)	Percentage	
		Recyclable	Non- recyclable
Food waste and yard waste	17.449	77.26	22.74
Plastic	3,096	70.23	29.77
Paper	1,605	56.21	43.79
Metal	0,314	0.00	100
Glass	1,046	95.69	0.31
Rubber	0,024	0.00	100

Textile	0,382	0.00	100
Wood	4,329	0.00	100

3.4. Recommendation for solid waste processing

In order to reduce waste transport into landfill, it is recommended to apply composting technology for food and yard waste, while other recyclable wastes such as plastic, paper, and glass, it can apply process of collection, separation, cleansing, and packing, before selling into nearby shanties. Furthermore, it will need special station or building such as recycling centre in the tourism area.

4. Conclusion

Theaveragesolidwastegenerationunit in theHarau Valley TourismAreais 3.053 Lit/cap./dayor 0.158 kg /cap./day. Thecomposition of wasteconsist of foodwaste (34.29%), paper (8.39%), plastic (16.43%), textiles (1.12%), rubber (0.07%), yardwaste (17.05%), wood (12.74%), glass (3.08%), non-ferrousmetals (0.92%) and others (5.91%). Recyclingpotentialforfood and yardwetwaste (77.26%), plastic (70.23%), paper (56.21%), and glass (95.69%). Forprocessingoptionitisrecommedatedtoapplycompostingmethodforcompostablewaste, whileforotherrecyclablewaste can applycollection, separation, clensing, and packingbeforesellingintonearbyshanties.
Forfurtherstudy, thisrecommendation can be used as thebasisforplanningthesolidwastemanagementsystemforinfrastructures and non technicalaspect of Harau Valley.

Acknowledgment

Publication of this article is supported by Engineering Faculty of UniversitasAndalas grant under contract No.042/UN.16.09D/PL/2020

References

[1] Regional Regulation of Lima Puluh Kota Regency No. 7 of 2012 concerning the Lima Puluh Kota Regional Spatial Plan
[2] Department of Culture of Youth Tourism and Sports in Lima Puluh Kota Regency, 2017
[3] Bhat RA, Nazir R, Ashraf S, Ali M., Bandh SA, and Kamili AN 2014 Waste Manage Res. **32** 165– 169
[4] Zorpas AA, Irene Voukkali I, and Loizia P 2015 Desalines Water Treat. **56** 1141-1149
[5] Arbulu I, Lozano J, Rey-Maquieira J 2016 Waste Manage. **51** 252-258
[6] Hoang MG, Fujiwara T, and Pham Phu ST 2017 Journal of JSCE. **5** 123-132
[7] Giurea R, Precazzini I, Ragazzi M, Achim MI, Cioca LJ, Conti F, Torretta V, and Rada EC 2018 Resources. **7** 1-12

IOP Conf. Series: Materials Science and Engineering **990** (2020) 012015 doi:10.1088/1757-899X/990/1/012015

A Comparison of Palm Fatty Acid Distillate (PFAD) Esterification using Sulphated Alumina versus Sulphuric Acid Catalyst

M Ulfah [1], Firdaus[1], E Sundari[1], E Praputri[1]
[1]Departmen Chemical Engineering, Universitas Bung Hatta, Padang, West Sumatera, Indonesia

E-mail: mariaulfah@bunghatta.ac.id

Abstract. The selection of raw materials is the cost-wise strategy for biodiesel to compete with diesel. The raw material used in the second generation of biodiesel production is non-edible oil which contains high free fatty acids (FFA). Palm Fatty Acid Distillate (PFAD); the by-product of crude palm oil (CPO) industry has a high potential as the amount is abundant in Indonesia. The objective of this study is to compare the catalytic activity of esterification reaction using self-synthesized sulphated alumina and a sulphuric acid catalyst against the reduction of FFA. The reaction was carried out in a three-necked flask, at a temperature of 60°C, 7 hours of reaction time and rotation speed of 400 rpm. The results showed that the maximum FFA conversion of PFAD esterification reactions over sulphated alumina was 84%, achieved at methanol to oil ratio of 1.7 (v/v), catalyst to PFAD ratio of 12 % (w/w) and reaction time of 7 hours. Meanwhile, the optimum conditions for PFAD esterification usingsulphuric acid catalyst are methanol to oil volume ratio of 1.5; catalyst to PFAD ratio of 7.4 wt.% and 3 hours of reaction time. 99.34% of FFA conversion was achieved with the homogeneous catalyst.

1. Introduction

The biodiesel produced from non-edible oils is said to be the second-generation biodiesel[1]. In Indonesia, the availability of these non-edible oil resources such as castor oil (*Ricinus communis*), polanga (*Calophylluminophyllum*), rubber seed oil (*Hevea brasiliensis*), sea mango seed oil (*Cerberamanghas L*), and palm fatty acid distillate (PFAD) is very abundant. Palm fatty acid distillate is a by-product of the palm oil industry (crude palm oil, CPO) with up to 3.25% of PFAD is yielded from every tonne of CPO processed. As the largest producer of palm oil, Indonesia produced 32.5 million tons of CPO in 2015 [2]. Thus, the PFAD generated was 1,056,250 tons. This suggests that the development of the biodiesel industry in Indonesia over PFAD as raw material is substantial. Currently, the majority of PFAD is exported to overseas market as the source of vitamin E in the cosmetics industry, animal feed and raw materials for the oleo-chemical industry. On the contrary, in Indonesia PFAD utilization has not been explored.

PFAD comprise mainly of free fatty acids (FFA) ranged from 85 to 100%. An acid-catalyzed esterification pre-treatment, both homogeneous and heterogeneous, is the most suitable initial step to lower the FFA content. Heterogeneous catalysts have always been preferred over homogeneous

[1] mariaulfah@bunghatta.ac.id

Content from this work may be used under the terms of the Creative Commons Attribution 3.0 licence. Any further distribution of this work must maintain attribution to the author(s) and the title of the work, journal citation and DOI.
Published under licence by IOP Publishing Ltd

catalysts due to their easy separation from products and reusability, lesser purification steps, resulting in less water consumption.

Few numbers of heterogeneous catalysts application in PFAD esterification have been reported. PFAD esterification in the presence of modified Zr-based catalysts, i.e. SO_4-ZrO_2, WO_3-ZrO_2, and TiO_2-ZrO_2 has investigated [3]. Among the three types of catalysts, SO_4-ZrO_2 catalyst generated the highest biodiesel yield of 93.7% which was carried out at 80°C for 2hours with 6:1 methanol to PFAD molar ratio and catalyst loading of 1 wt.%. A similar investigation was reported [4]. It was concluded that the maximum methyl ester production of 81% was carried out in the presence of WO_3-ZrO_2 at 250°C for 1 minute. Furthermore, the study on mixed Cr : W : Ti oxides catalysts with different mass ratio has been studied[5]. The highest methyl ester content of 83% was obtained at a mass ratio of Cr : W : Ti = 1:1:1 using 2:1 methanol/PFAD molar ratio and increase in temperature until 170 °C for 3 hours. Another PFAD esterification by Shuit and Tan [6] using a multi-walled carbon nanotubes catalyst in a pressurized reactor was performed. The reaction was conducted at 170°C for 3 hours, methanol to PFAD mole ratioof 20 and catalyst loading of 2 wt.%. It was found that the multi-walled carbon nanotubes catalyst prepared via the inset polymerization of poly (sodium4-styrenesulphonate) gave a biodiesel yield of 93.4%. Moreover, Olutoye *et al.* compared the performance of three different solid acid catalysts (ZrFeTiO, ZrFeO and FeTiO) prepared using the sol-gel method [7]. The highest PFAD conversion of 96.45% was observed for the ZrFeTiO catalyst at a temperature of 170°C, 5 hours of reaction time, methanol to PFAD molar ratio of 3:1 and 3 wt.% catalyst loading. The activities of $H_3PW_{12}O_{40}$ and $Cs_xH_{3-x}PW_{12}O_{40}$ supported on SiO_2, MCM-41 and ZrO_2 have been studied[8]. The highest reactivity and stability of the catalyst was obtained using $Cs_{1.5}H_{1.5}PW_{12}O_{40}$ supported with 20 wt.% MCM-41. Methyl ester obtained was 92% at temperature of85°C, 4 hours of reaction time, 12wt.% catalyst amount and methanol to PFAD molar ratio of 15. The polymeric mesoporous SO_3H-$ZnAl_2O_4$ catalyst prepared at a sulfonation temperature of 100°C and the molar ratio of Al/(Zn + Al) equal to 1 has reported [9]. Resulted in 94.65% of methyl ester obtained [9]. This biodiesel recovery is achieved at a reaction temperature of 100°C, reaction time of 60 min, methanol to PFAD molar ratio of 9:1, and catalyst concentration of 1 wt.%.

In light of the foregoing, the application of sulphated alumina as a catalyst in the esterification of PFAD has not been well explored. The performance of sulphated alumina catalyst in esterification of rubber seed oil and used cooking oil has studied [10]. As reported by this research, the sulphated alumina catalyst has a large specific surface area and pore size in addition to high thermal stability. The maximum conversion of 97% rubber seed oil was achieved under the optimum reaction conditions; methanol to oil volume ratio of 1.8; concentration of catalyst/oil of 1.67% (w/v); 7 hours of reaction time and a temperature of 70°C. Meanwhile, 89.97% of used cooking oil conversion was obtained at the optimum reaction conditions; methanol to oil volume ratio of 1.5; catalyst/oil concentration of 1% (w/v); 1 hour of reaction time and temperature of 60°C [11]. Accordingly, the objective of this research was to study the activity of self-synthesized sulphated alumina catalyst, the optimum operating conditions (ratio of methanol to PFAD and catalyst loading)in addition to performance comparison of the catalyst to homogeneous sulphuric acid catalyst.

2. Materials and Method

2.1 Materials

Palm fatty acid distillate (PFAD) used in this research was obtained from PT. Incasi Raya-Padang which contains 95.23 wt.% of free fatty acids (FFA). The chemicals used were including boehmite (from PT. Pertamina-Indonesia), HNO_3 (67%, Merck), NH_3 (25%, Merck), in addition to sulphuric acid (precursor of sulphated alumina catalyst) and methanol used for esterification reaction (technical grade) supplied by Bratachem, Inc., Indonesia. Meanwhile, alcohol (99%, Merck); phenolphthalein (Merck), and sodium hydroxide (technical grade) were used in FFA analysis.

IOP Conf. Series: Materials Science and Engineering **990** (2020) 012015 doi:10.1088/1757-899X/990/1/012015

2.2. Catalyst synthesis

Sulphated alumina catalyst synthesis has been reported [10]. The catalyst was prepared in two stages; gamma-alumina catalyst synthesis followed by wet impregnation with concentrated sulphuric acid.

2.3. Esterification reaction

The esterification reaction was conducted in a batch-wise system using 500 mL three-necked glass flask equipped with hot plate, magnetic stirrer, thermometer and coupled to a condenser to avoid alcohol loss. In all experiments, approximately 25 g (30 mL) of PFAD was weighed and heated at 60°C. Different methanol to PFAD volume ratio in the range of 1.1:1 to 1.7:1, and catalyst loading (12.0 wt.% to 28.0 wt.% of PFAD) were used to investigate their influences on esterification reaction of PFAD. After heated, methanol and catalyst were added into melted PFAD. Following completion of the reactions, the mixture was cooled, discharged and centrifuged. The FFA conversion was determined by the following (AOCS 5a-40) standard method as Eq. (1), where AV_f and AV_p were assigned for the acid value of feedstock and product, respectively.

$$\%FFA = \frac{AV_f - AV_p}{AV_f} \times 100\% \tag{1}$$

3. Results and Discussions

Changes in textural properties (i.e. surface area, pore diameter and pore volume) of alumina caused by impregnation of concentrated sulphuric acid into γ-Al_2O_3 for 3 hours have been reported. The pore diameter increases, but the surface area and pore volume of the sulphated alumina decreases with the increase of sulfate loading [11]. The surface area, pore volume and pore diameter of self-synthesized catalyst were 76.387 m^2/g; 0.11 cm^3/g and 63.64 Å, respectively.

In addition to the type of catalyst, other factor affecting the FFA conversion are the ratio of methanol to raw materials, catalyst loading, reaction time and temperature along with the stirring speed. In this study, the reaction temperature was set at 60°C and stirrer speed of 400 rpm. Earlier research has shown that methyl ester generated from PFAD esterification reaction was obtained at 7 hours of reaction time.

3.1. PFAD esterification using self-synthesized sulphated alumina solid catalyst

3.1.1. Effect of methanol to oil volume ratio

Catalyst loading used in this experiment was 12 wt.%. Figure 1 depicts a significant increase of FFA conversion with respect to methanol/PFAD volume ratio ranged between 1: 1.3 to 1: 1.7.

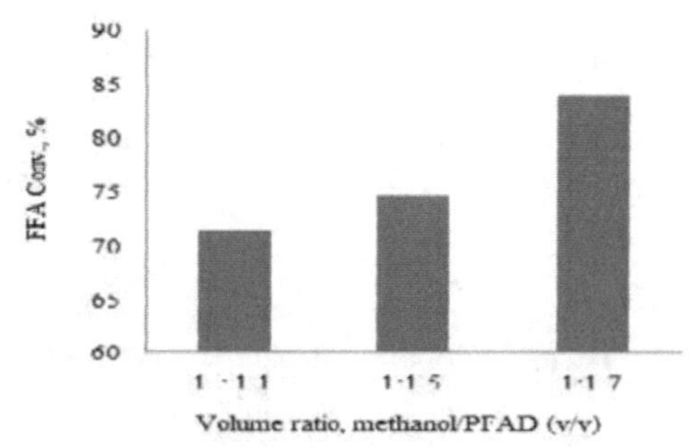

Figure 1. Effect of methanol to PFAD volume ratio on FFA conversion (T = 60°C, catalyst loading 12 wt.% and 7 hours)

An excess amount of methanol as reactant was added to shift the reaction towards the product, in principle, because esterification is a reversible reaction, according to the following stoichiometric equation.

$$R-COOH + R'{-}OH \leftrightarrow R{-}CO{-}CH2{-}R' + H2O \quad (2)$$

However, this method is not effective where higher methanol to oil ratio will no longer increase FFA conversion [12]. This lack of a significant increase in the esterification reaction, which thus reacted with the PFAD and, therefore pushed the reaction backwards. In this study, the optimum methanol/oil volume ratio was 1.7 (shown in Figure 2).

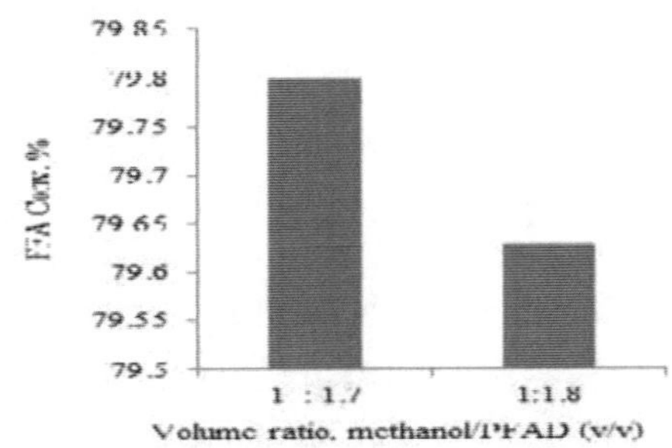

Figure 2. Effect of methanol to PFAD volume ratio on FFA conversion (T = 60°C, catalyst loading 28 wt.% and 7 hours)

3.1.2. Effect of catalyst loading

The effect of sulphated alumina catalyst loading on FFA conversion was also investigated (Figure 3). In contrast to the amount of methanol, FFA conversion decreases with the increase of sulphated alumina catalyst loading. FFA conversion of 84% was obtained when methanol/PFAD volume ratio of 1:1.7 and sulphated alumina catalyst loading of 12 wt.% were used. Nevertheless, the conversion was decreased by approximately 3.6% at catalyst loading of 20 wt.%. These results demonstrated PFAD esterification reactions limited by mass-transfer in addition to the equilibrium that has been established [13].

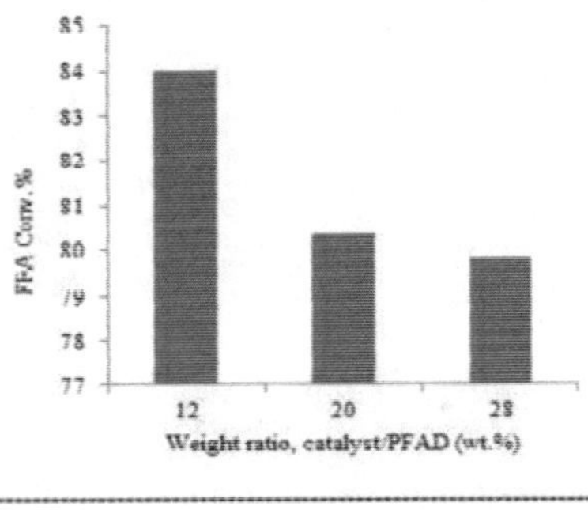

Figure 3. Effect of catalyst loading on FFA conversion (volume ratio methanol to PFAD of 1.7, T = 60°C, and 7 hours)

3.2. PFAD esterification using homogeneous sulphuric acid catalyst

3.2.1. Effect of methanol to oil volume ratio

This study also investigated the catalytic activity of PFAD esterification using a sulphuric acid homogeneous catalyst. The effect of the methanol/PFAD volume ratio on FFA conversion was observed in the range of 1: 0.9 to 1: 1.5. The reaction was conducted at 60°C with the amount of sulphuric acid catalyst of 7.4 wt.%, magnetic rotation speed of 400 rpm and 3 hours of reaction time. As the amount of methanol in the PFAD esterifications reaction increases, the FFA conversion also increases as shown in Figure 4. FFA conversion of more than 99% was achieved at methanol to PFAD volume ratio of 1.5 and the next experiments will be carried out under these conditions.

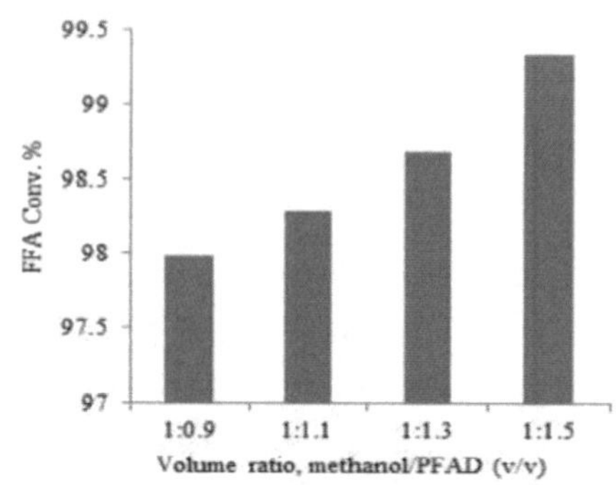

Figure 4. Effect of methanol to PFAD volume ratio on FFA conversion (T = 60°C, catalyst loading 7.4 wt.% and 3 hours)

3.2.2. Effect of catalyst loading

25 g of PFAD was used in this study and after melting at 40°C the volume obtained was 30 ml. the volume of sulphuric acid observed was 0.7; 1.0 and 1.3 ml (the density of concentrated sulphuric acid is assumed to be 1.84 gml^{-1}),thus the catalyst/PFAD weight ratio is 5.2%; 7.4% and 9.6%. From Figure 5 it is concluded that an appropriate amount of sulphuric acid catalyst is needed to achieve optimum FFA conversion. The optimum FFA conversion was achieved at sulphuric acid loading of 7.4 wt.%. Increasing the sulphuric acid catalyst loading to 9.6 wt.% did not significantly increase the PFAD esterification reaction. This result showed that the equilibrium has been reached.

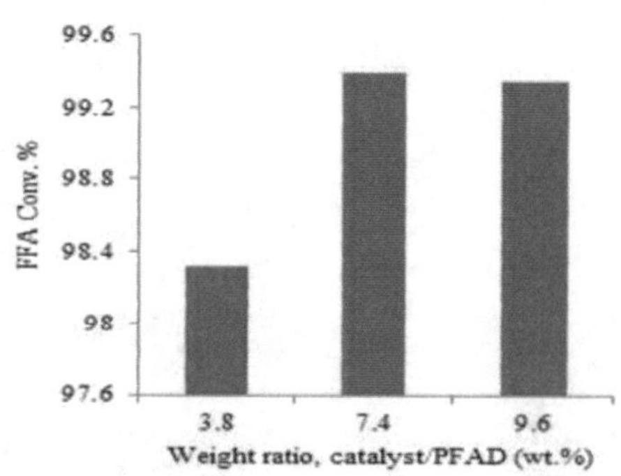

Figure 5. Effect of catalyst loading on FFA conversion (volume ratio methanol to PFAD of 1.5, T = 60°C and 7 hours)

3.2.3. *Effect of reaction time*

Optimum reaction time was essential to achieve maximum FFA conversion allowing sufficient contact between catalyst, methanol and PFAD. Figure 6 showed that FFA conversion increase on 1–3 hours. Several investigations [14,15] concluded if the reaction time is set longer, the reaction can shift toward the reactants because the esterification reaction is a reversible reaction.

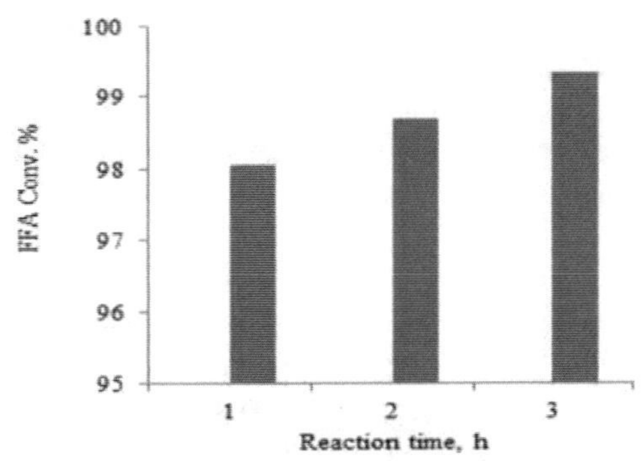

Figure 6. Effect of reaction time on FFA conversion (methanol to PFAD volume ratio of 1.5,T = 60°C and catalyst loading 7.4 wt.%)

A side from high conversion of FFA catalyzed using sulphuric acid, the acid condition of reaction can cause corrosion of equipment. Furthermore, after reaction was completed, the mixture resulting from the esterification reaction need to be washed to avoid environmental contamination. Accordingly, the use of sulphuric acid homogeneous catalysts was shifted to the acid heterogeneous catalysts [3-6]. As stated previously in the introduction, the optimum temperature for the conversion of FFA from PFAD using a solid catalyst over 90% was achieved at temperatures above 60 °C. In this study, the reaction was carried out at a temperature of 60 °C.

4. Conclusions

The optimum conditions for the esterification reaction catalyzed by sulphated alumina and sulphuric acid were different, and so does the FFA conversion achieved. Heterogeneous sulphated alumina catalyst requires higher optimum parameters than the sulphuric acid catalyst, but allowing lower product generated. The maximum FFA conversion of 84% was obtained using sulphated alumina catalyst at methanol to PFAD volume ratio of 1.7, catalyst loading of 12 wt.% and 7 hours of reaction time. Meanwhile, the optimum parameters for PFAD esterification using a sulphuric acid catalyst are volume ratio of methanol/PFAD volume ratio at 1:1.5 with 7.4 wt.% catalyst loading and 3 hours giving a 99.34% reduction in FFA.

Acknowledgement
Authors would like to thank Riza Kurnia Putri and Widya Oktari for the help during the research.

References

[1] Bikkavolu J and Swarna K 2019 *Recent Advances in Material Sciences* 693-709
[2] Nazratul Z A K, Gaanty P M, Mohd HAR and Mashitah MY 2017 *J Clean Prod* 143 1-9
[3] Mongkolbovornkij P, ChampredaV, SutthisripokW and Laosiripojana N 2010 *Fuel Process.* Technol.911510-16
[4] Laosiripojana N, Kiatkittipong W, Sutthisripok W andAssabumrungrat S2010 *Bioresour. Technol.*1018416-23
[5] Wan Z and Hameed B H 2014 *Energy Convers. Manag.*88 669-676

[6] Shuit S H and Tan S H 2014 *Energy Convers. Manag.*88 1283-89
[7] Olutoye M A, Wong C P, Chin L H and Hameed B H 2014 *Fuel Process. Technol.* 124 54-60
[8] Trakarnpruk W 2013 *Mendeleev Commun.* 23 46-48
[9] Soltani S, Rashid U, Yunus R and Taufiq-Yap Y H 2016 *Fuel*178 253-262
[10] Ulfah M and Subago 2016 *Reaktor*16(3) 109-115
[11] Ulfah M, Firdaus, Octavia S, Suherman H and Subagjo 2018 *IOP Conf. Ser. Mater. Sci. Eng.* 543:012007
[12] KonwarL J, Das R, Thakur AJ, SalminenE, Mäki-Arvela P, Kumar N., Mikkola, J-P and Deka D 2014 *J Mol.Catal. A: Chem.*388–389 167–176
[13] Yujaroen D, Motonobu G, Sasaki M and Shotipruk A 2009 *Fuel* 88(10) 2011-16
[14] Akinfalabi S-I, Rashid U, Yunus R and Taufiq-Yap Y H 2017 *Renew. Energy* 111 611-619
[15] Syazwani O N, Rashid U, Mastuli M S and Taufiq-Yap Y H 2019 *Renew. Energy* 131 187-196

IOP Conf. Series: Materials Science and Engineering **990** (2020) 012016 doi:10.1088/1757-899X/990/1/012016

WaterDistribution in the Flow Field of the BipolarPlate on the CathodeSide of the PEM FuelCell

Mulyazmi[1], M Marthynis[1], F R Safitri[1], N G Sari[1]
[1]Department of Chemical Engineering, Universitas Bung Hatta
Padang 25586, Indonesia

E-mail: mulyazmi@yahoo.com

Abstract. The presence of water in the flow field greatly affects the performance of PEM fuel cell. One of the problems is caused by the water imbalance in the PEM fuel cell. If the water content is too high it causes flooding in the PEM fuel cell stack. As a result, the distribution flow of raw materials (hydrogen and oxygen) flowing into the stack is blocked. Conversely, if it is too dry it will cause the membrane to crack. Both cases resulted in a decrease in the performance of the PEM fuel cell system. The method used for water management in the flow field of a bipolar plate flow Computating Fluids Dynamics (CFD) software. The results showed that the highest water content occurred in the spiral flow field at a flow rate of 4 x 10^{-8} kg /s with a water content of 0.75 g /g. The lowest water content occurs in the multiple channel serpentine flow field at a flow rate of 4 x 10^{-8} kg / s with the water content of 0.47 g / g. In this condition the performance of PEM fuel cell is relatively stable in the single channel serpentine flow field with a flow rate of 4 x 10-8 kg / s and a depth of 0.3 mm because the formation of water is insignificant compared to using other flow fields.

1. Introduction

One of the alternative energy sources that are environmentally friendly is PEM fuel cell[1] and from biomass such as palm oil mill effluent [2,3]. This energy source uses hydrogen and oxygen (air) as raw materials. PEM Fuel Cell technology has properties such as: a clean, safe energy source, high current density and efficiency, and low operating temperature[1].

The water content in the flow on the PEM fuel cell stack has an influence on the performance of the PEM fuel cell. The presence of water content in the stack has a significant effect on the performance of the PEM Fuel Cell system. The water source in the PEM fuel cell stack is: the result of the electrochemical reaction between hydrogen and oxygen, and the water flowing with the reactants enters the cathode and anode sides. According, the water content in the flow field at the cathode side has more effect on achieving optimum performance than the water content on the anode side. The main function of the flow field in the PEM Fuel Cell system is to distribute the total reactant gas on the *Gas Diffusion Layer* (GDL) surfaceand for the outflow of water as a result of the reaction from the fuel cell stack[1]. An important problem in the operation of the PEM Fuel Cell system is water management. Water balance in the PEM Fuel Cell system includes control of the water content that

[1]mulyazmi@yahoo.com

Content from this work may be used under the terms of the Creative Commons Attribution 3.0 licence. Any further distribution of this work must maintain attribution to the author(s) and the title of the work, journal citation and DOI.
Published under licence by IOP Publishing Ltd

can be in the system. This condition is carried out in a channel-shaped flow field found on a bipolar plate that has several patterns or structures.

Several studies related to water management to improve the performance of the PEM Fuel Cell system include by analyzing three different serpentine type flow field models. This study makes a difference to the ratio of the channel width or the rib width and different pressure loss[4]. By showing the effect of operating conditions on the water content flowing on the cathode side. The highest water content occurs at a current density of 0.9, with a relative humidity at the anode and cathode of 90%. Under these conditions, the system performance is relatively stable at 0.67 volts[5].PEM fuel cell performance was more influenced by the increase in relative humidity on the cathode side compared to the increase in relative humidity on the anode side[6]. The mass fraction of oxygen and hydrogen decreases along the flow direction and the amount of water vapor increases along the flow direction on the anode and cathode sides[7].

Flow field design is one of the factors that have a significant effect on the distribution of reactant gas. This research focuses on the distribution of water in the flow field of the Bipolar Plate on the Cathode Side of the PEM Fuel Cell, including: straight parallel, spiral, single channel serpentine and multiple channel serpentine. The purpose of this study is to compare the type of flow field configuration model with the variation in channel depth and velocity of the reactant flow into the fuel cell stack. identification of water formation in the flow field in the cathode side channel using Computating Fluids Dynamics(CFD) Software.

2. Methodology

The function of the PEM Fuel Cell stack is to react hydrogen with oxygen to produce an electric current and water. The process design of the fuel cell system is shown in Figure 1

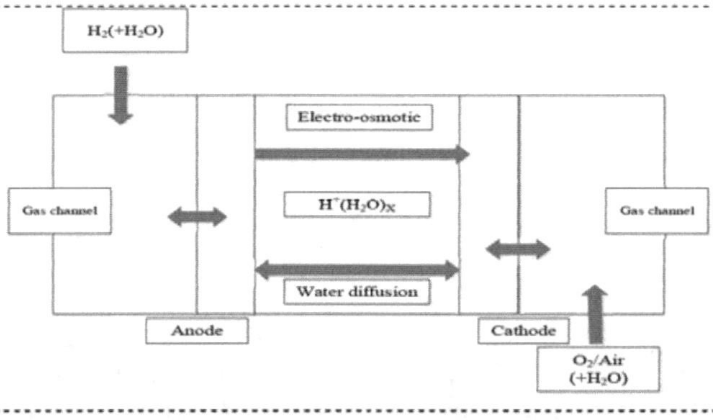

Figure 1. Single Cell Process Design Diagram in PEM Fuel Cell System

The modeling design uses the one-phase assumption as laminar flow. The effect of flow field is shown by varying the flow rates of hydrogen and oxygen as reactant gases and different channel depths. The parameters used are: cell temperature is 353,15 K, pressure is 101352 pa, membrane thickness is 0,06 mm, catalyst layer thickness is 0.03 mm, diffusion layer thickness is 0,3 mm, MPL layer thickness is 0,12 mm,CC layer thickness is 0,6 mm, Mass ratio (H_2:H_2O) is 0,6 : 0,4, Mass ratio (O_2;H_2O), relative humidity is 75%. Type of flow field in PEMFC system design is shown in Figure 2

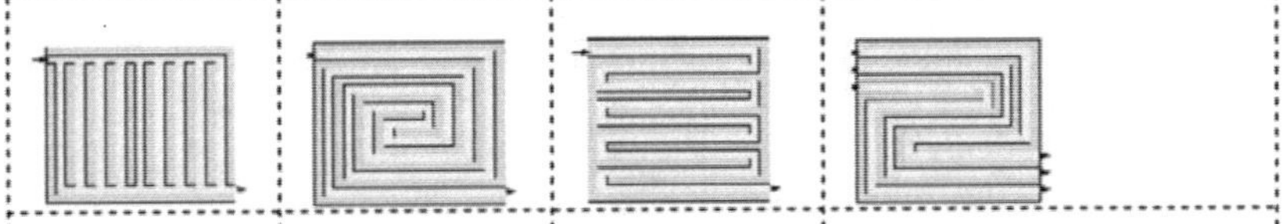

Figure2.Flow Field Type Configuration Schematic;(a)straight parallel (b)spiral (c)single channel serpentine (d)multiple channel (triple) serpentine[8]

Generally the channel arrangement can be: direct flow, countercurrent flow, or cross flow.

3. Mathematical Models

A. Flow rate of reactants

Mass flow rate of the reactantson the PEM fuel cell stack are as follow:

$$\dot{m}_{H2} = \frac{I}{2F} M_{H2} \quad (1)$$

$$\dot{m}_{O2} = \frac{I}{4F} M_{O2} \quad (2)$$

Mass flow rate of the water produced in the PEM fuel cell stack is as follows

$$\dot{m}_{H2O} = \frac{I}{2F} M_{H2O} \quad (3)$$

The molar volume is:

$$v_m = \frac{V}{N} = \frac{RT}{P} \quad (4)$$

Under standard conditions (atmospheric pressure and 15 ° C) the molar volume is:

$$v_m = \frac{RT}{P} = \frac{8.314 \times 288.15}{101,300} = 0.02365\,m^3\,mol^{-1}$$
$$= 23.651\,mol^{-1} = 23,650\,cm^3\,mol^{-1} \quad (5)$$

The volumetric flow rate (standard liters per minute) slpm is:

$$\dot{V}_{H2} = 23.65 \times 60 \times \frac{I}{2F} \quad (6)$$

$$\dot{V}_{O2} = 23.65 \times 60 \times \frac{I}{4F} \quad (7)$$

B. Flow Rate in the Flow field

The flow velocity in the flow area is influenced by the hydraulic diameter, channel length and flow rate. The hydraulic diameter D_H is described by the following equation:

$$D_H = \frac{2 \times w_c \times d_c}{w_c + d_c} \quad (8)$$

Where w_c is the channel width and d_c is the channel depth. The flow rates of the components flowing into the PEM fuel cell stack are:

$$Q_{stack} = \frac{I}{4 \times F} \times \left(\frac{S}{r_{O_2}}\right) \times \frac{R \times T}{P_{in} - \varphi \times P_{sat}} \times N_{cell}. \quad (9)$$

I iscurrent density, Faraday's constant is oxygen stochiometric ratio , $ro_2$is oxygen content in the air, T is temperatur , Pin is the inlet pressure of the fuel cell stack, Φ is relative humidity, Psat issaturation pressure, The rate of velocity flowing into the fuel cell stack

$$\bar{V} = \frac{Q_{stack}}{N_{celi} \times N_{ch} x \; Wc \; x \; dc} \quad (10)$$

Ncell is number of cells, Nch is the number of parallel channels in the fuel cell stack, wc is channel width and dc is channel depth.

4. RESULT AND DISCUSSION

This design shows the distribution of water in the flow field along the channel and also to determine the formation of stagnant areas on the cathode surface.

4.1 Effect of Flow Rate of Reactants (O_2) and Channel Depth on Distribution of Water Formation

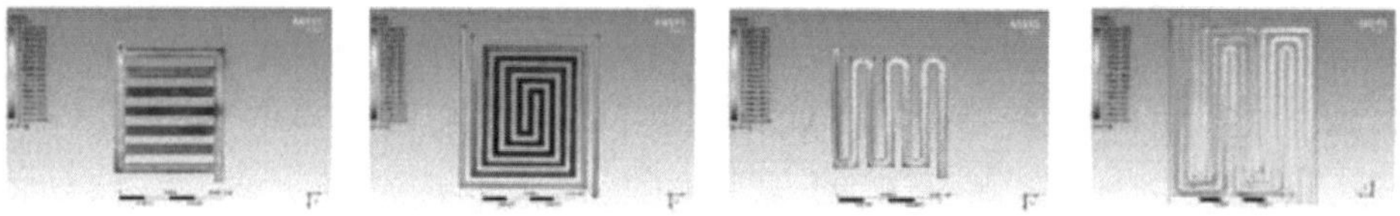

(a) (b) (c) (d)

Figure 3. Flow rate of reactants is 4 x 10^{-7} kg/s and the channel depth is 0,3 mm; (a) Straight Parallel, (b) Spiral (c) Single channel serpentine (d) multiple channel

Figure 3a shows the uneven distribution of water in the flow plane along the cathode side channel. The inflow indicates flooding and the inside channel is dry, while the outflow shows a stable water distribution. Figures 3c and 3d show a large amount of water accumulating in the channel. This condition shows that the pressure is very dominant to distribute water to the outlet flow of the channel and also to facilitate oxygen transport[9].

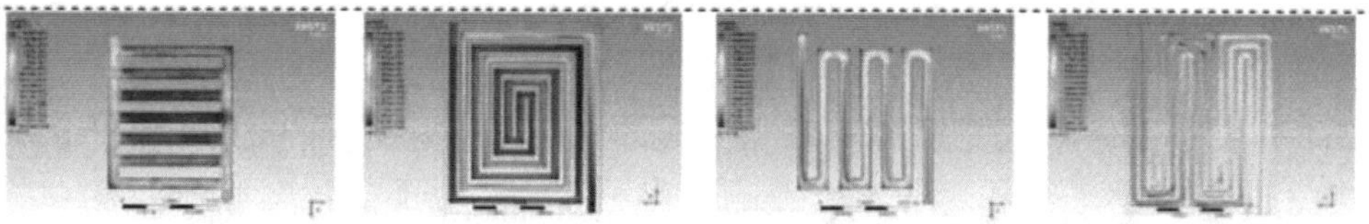

(a) (b) (c) (d)

Figure 4. Flow rate of reactants is 7 x 10^{-7} kg/s and the channel depth is 0,3 mm; (a) StraightParallel, (b) Spiral (c) Single channel serpentine (d) multiple channel

Figure 4a shows the flow of inlet and outlet on the channel is flooded and inside the channel there is dryness. Figures 4c and 4 show the distribution of water tends to be flooded. this condition results in a decrease in fuel cell performance[10].

IOP Conf. Series: Materials Science and Engineering **990** (2020) 012016 doi:10.1088/1757-899X/990/1/012016

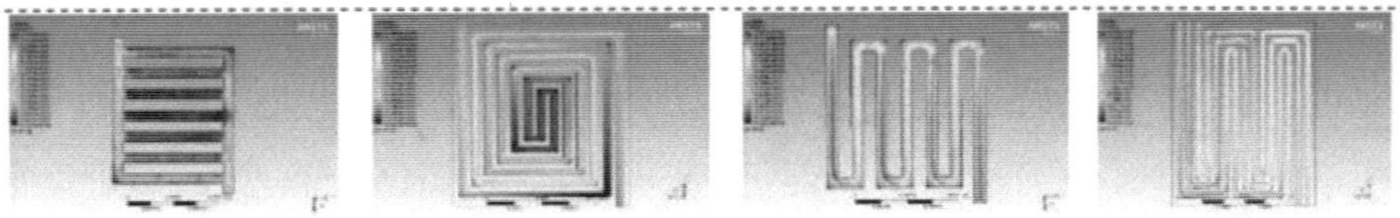

(a) (b) (c) (d)

Figure 5 Flow rate of reactants is 9 x 10^{-7} kg/sand the channel depth is 0,3 mm; (a) Straight Parallel, (b) Spiral (c) Single channel serpentine (d) multiple channel

Figure 5a shows the inlet, outlet flooding and dryness on the inside of the channel. Figure 4.3b shows the dryness at the inlet, outlet and inside of the channel. Figures 5c and 5d tend to be flooded as a result of poor gas flow distribution and high water content resulting in decreased cell performance [10].

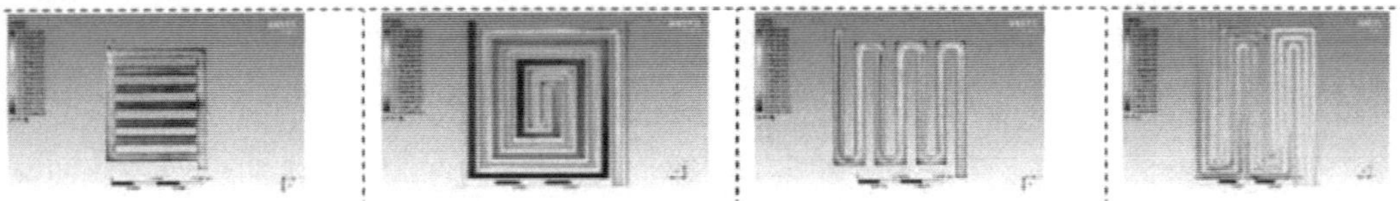

(a) (b) © (d)

Figure 6 Flow rate of reactants is4 x 10^{-7} kg/sand the channel depth is 0,5 mm; (a) Straight Parallel, (b) Spiral (c) Single channel serpentine (d) multiple channel

Figure 6a shows the flow of inlet and outlet on the channel is flooded and inside the channel there is dryness. Figures 6c and 6d show the distribution of water tends to be flooded.

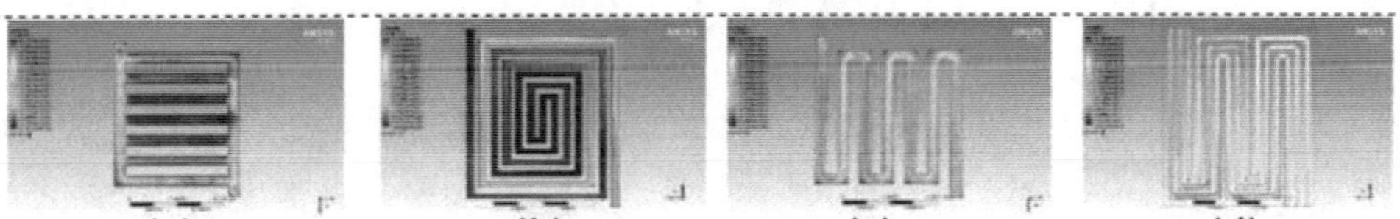

(a) (b) (c) (d)

Figure 7 Flow rate of reactants is7 x 10^{-7} kg/sand the channel depth is 0,5 mm; (a) Straight Parallel, (b) Spiral (c) Single channel serpentine (d) multiple channel

Figure 7a shows the flow of inlet and outlet on the channel is flooded and inside the channel there is dryness.Figures 7c and 7d show the distribution of water tends to be flooded. this condition results in a decrease in fuel cell performance[10].

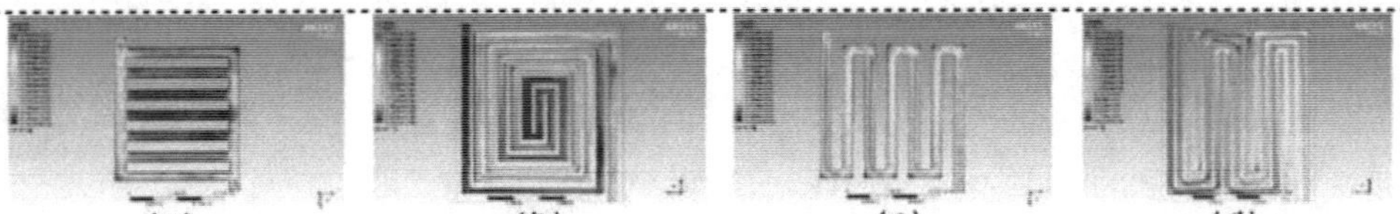

(a) (b) (c) (d)

Figure 8 Flow rate of reactants is9 x 10^{-7} kg/sand the channel depth is 0,5 mm; (a) Straight Parallel, (b) Spiral (c) Single channel serpentine (d) multiple channel

Figure 8a shows the flow of inlet and outlet on the channel is flooded and inside the channel there is dryness.Figures 8c and 8d show the distribution of water tends to be flooded. this condition results in a decrease in fuel cell performance[10].

4.2 Comparison of Water Formation Distribution with Flow Rate Variations

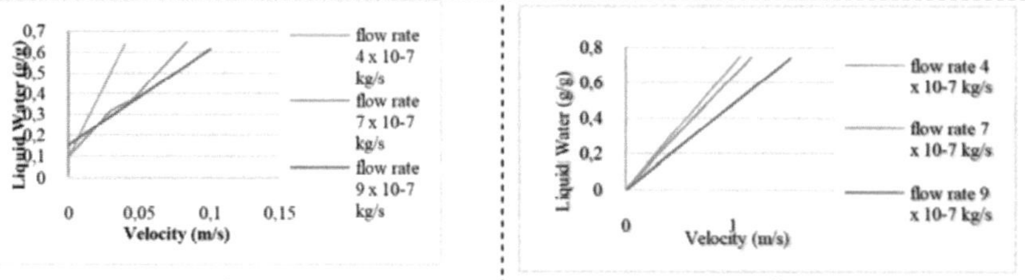

(a) (b)

Figure 9 Comparison Graph of Water Formation in; (a) Straight Parallel (b) Spiral

In Figure 9a, shows that at a flow rate of 4x10^{-7}kg/s, 7x10^{-7}kg/s, and 9x10^{-7}kg/s the maximum amount of water is 0.63 g/g, 0.64 g/g, and 0.61 g/g. In Figure 10b shows that at a flow rate of 4x10^{-7}kg/s, 7x10^{-7}kg/s, and 9x 10^{-7}kg/s the maximum amount of water is 0.75 g/g, 0.74 g/g, and 0.73 g/g.Theoretically the maximum water content at the cathode is 1.4 g water/g air. Increasing the flow rate results in decreased water formation on the cathode side. This is due to the spiral flow field design resulting in inefficient water formation which can reduce cell performance[11].

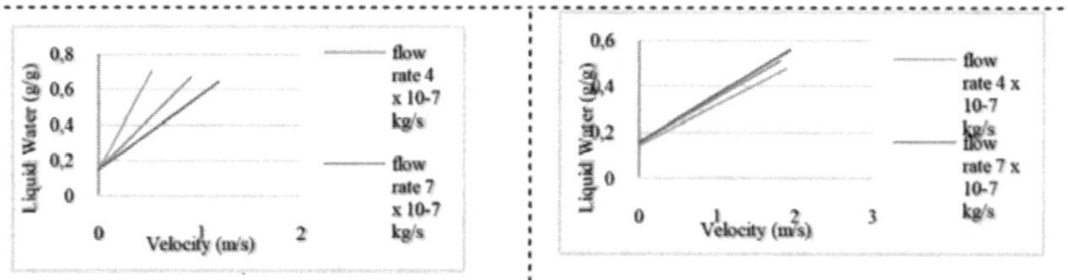

(a) (b)

Figure 10. Comparison Graph of Water Formation in; (a) Single Channel Serpentine (b) Multiple Channel Serpentine

In Figure 10ashows that at a flow rate of 4 x 10^{-7}kg/s, 7 x 10^{-7}kg/s, and 9 x 10^{-7}kg/s the maximum amount of water is 0.70 g/g, 0.66 g/g, and 0.64 g/g.In Figure 10bshows that at a flow rate of 4 x 10^{-7}kg/s, 7 x 10^{-7}kg/s, and 9 x 10^{-7}kg/s the maximum amount of water is 0.47. g/g, 0.50 g/g, and 0.55 g/g. High amounts of water can reduce PEMFC performance. Whereas the multiple channel serpentine flow field produces a low amount of water, a low amount of water can improve PEMFC performance. This shows that the flow field design is an important issue in improving PEMFC performance[12].

4.3 Validation

In this study the results of numerical simulations are compared with the result of experiments conducted by Zinko, as show in Figure 11.

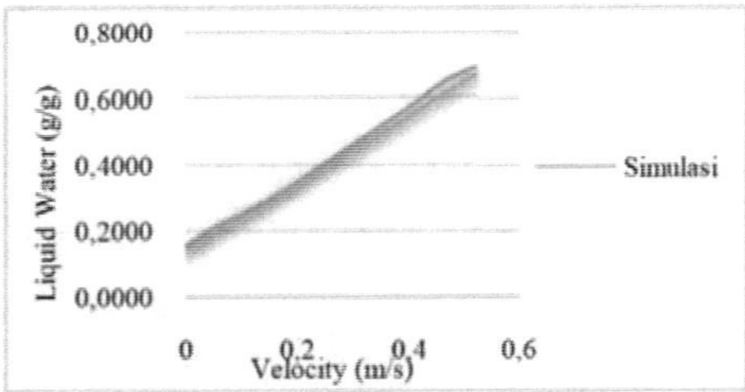

Figure 11 Validation of simulation results with experiment

In Figure 11, it can be seen that this prediction has the same tendency as the experimental data in terms of the amount of water formed in the Single Channel Serpentine type flow field design, because this type of channel can distribute water better than other channels.

5. CONCLUTION

From the research it can be concluded:

1. The type of flow field that is in the safe zone or in good condition is a single serpentine channel with a flow rate of 4×10^{-7}kg/s and a depth of 0.3 mm
2. The difference in the performance of these different flow planes also comes from the amount of air accumulated and the deeper channels.
3. The validation results show a similar trend.

References

[1]. Mulyazmi2017 International Journal Advanced Science Engineering Information Technology**7** 2

[2]. Hazmi, A., Desmiarti, R., Waldi E.P, Emeraldi, P 2016, Journal of Engineering and Technological Sciences, 48(1), 21-30.

[3]. Hazmi, A., Desmiarti, R., Waldi E.P, Eduardo, Emeraldi, 2017 Telkomnika (Telecomunication Computing Electronics and Control), 15(2) 554-559.

[4]. Ionescu, V 2019 Journal Of Physics: Conf.Series**1297** 9

[5]. Mulyazmi, Daud, W. R., Rahman, E. D., Purwantika, Muya, P. A., and Sari, N. G 2019International Journal Of Technology**10**634-643

[6]. Mulyazmi, Daud, W., Octavia, S., and Ulfah, M 2018 MATBC Web of Conferences **156** 6

[7]. Zinko, T., Pianko-Oprych, P., and Jaworski, Z 2018 Chemichal and Process Engineering 2018 **39** 143-154, 152

[8]. Cao, C., LI, Z. B., Wang, X. L., Zhao, X. B., and Han, W. Q 2014 Front. Energy Res.Retrievedfrom https://doi.org/10.3389/fenrg.2014.00025

[9]. Wang, C., Zhang, Q., Shen, S., Yan, X., Zhu, F., Cheng, X., and Zhang, J2017Scientific Reports. doi:10.1038/srep43447

[10]. Fahim, K. H., Alfayydh, E. M., and Dhahad, H. A2016World Research And Convection On Engineering Ang Technology1

[11]. Heinzel, A., Mahlendorf, F., and Jensen, C 2009*Elsevier. B. V*, 812. doi:10.1016/B978-044452745-5.00226-4

[12]. Barbir, F., 2012PEM Fuel Cells Theory and Practice.*Split: UNIDO-International Centre for Hydrogen Energy Technologies.*

IOP Conf. Series: Materials Science and Engineering **990** (2020) 012017 doi:10.1088/1757-899X/990/1/012017

Linear and Non-linear Regression Analysis for the Kinetics of $[AuCl_4]^-$ Removal by *p*-Hydroxybenzoate intercalated Mg/Al-Hydrotalcite

L A Hasnowo[1], S J Santosa[2], B Rusdiarso[2]
[1]Department of Nuclear Chemical Engineering, Polytechnic Institute of Nuclear Technology, Yogyakarta, Indonesia
[2] Department of Chemistry, Faculty of Mathematic and Natural Science, UniversitasGadjahMada, Yogyakarta, Indonesia

E-mail: lutfi.aditya@batan.go.id

Abstract. The best fit equations of linear and non-linear forms of the two widely used kinetic models, namely pseudo-first-order (PFO) and pseudo-second-order (PSO) equations, were compared in this study. The experimental kinetics of $[AuCl_4]^-$removal by*p*-hydroxybenzoate intercalated Mg/Al-hydrotalcite was used for this research. R^2 coefficient was employed as error analysis methods to determine the best-fitting equations. The results show that the non-linear form was better at best-fitting than the linear form. The experimental kinetics may have been distorted by linearization of the linear kinetic equations, and thus, the non-linear forms of kinetic equations should be primarily used to obtain the adsorption parameters.

Introduction

Sorption processes are proved to be an effective method for uptake of metal ions from solution [1-3]. hydrotalcite potentially have high capability for adsorption. hydrotalcite adsorption ability is supported by their high anion exchange and surface area [4,5]. In order to enhance the effectiveness of their adsorption capability, many researchers have modified hydrotalcite by various ways. There are some hydrotalcite modification techniques, but surfaces and galleries modifications were more often conducted. Many different anions can be intercalated into the interlayer space. In order to enhance the effectiveness of their adsorption capability, *p*-hydroxybenzoate was chosen as modifier for hydrotalcite [6].

The nature of sorption process will depend on physical or chemical characteristics of the adsorbent systems and also on the system conditions. Adsorption kinetics is an important parameter to describe the characteristics of an interaction between adsorbent and adsorb ate, which in turn controls the resident time in the adsorbent-solution interface. The accuracy of the model for estimating it is generally a function of a number of independent parameters. The mathematical simplicity of applying the model and its wide use in various data adsorption will have implications for its popularity.

In recent years, the linear forms of the PFO and PSO equations were widely used to determine the most fitted kinetic model for the adsorption process. When using the linear form, experimental adsorption kinetics should be linearized for the linear least-squares regression to estimate the model

[1] lutfi.aditya@batan.go.id.

Content from this work may be used under the terms of the Creative Commons Attribution 3.0 licence. Any further distribution of this work must maintain attribution to the author(s) and the title of the work, journal citation and DOI.
Published under licence by IOP Publishing Ltd

IOP Conf. Series: Materials Science and Engineering **990** (2020) 012017 doi:10.1088/1757-899X/990/1/012017

parameters. It has been reported that transformations of non-linear equations to linear forms implicitly alter their error structure in the measurement of model parameters [7-11]. As a result, one may obtain different kinetic parameters when using different forms of model equations for a given adsorption process. However, recently a nonlinear optimization model has been developed for the adsorption kinetics model. A number of researchers have conducted studies on the accuracy of the application of linear and non-linear regression models to explain the adsorption kinetics parameters.

Currently, however, there are few studies comparing the linear and non-linear forms of the PFO and PSO models in estimating the kinetic parameters for the adsorption of metal ions on hydrotalcite. Thus, in this study, the non-linear forms of the PFO and PSO equations were used to fit the experimental kinetic data for the adsorption of $[AuCl_4]^-$onto *p*-hydroxybenzoate intercalated Mg/Al-hydrotalcite, and a comparative analysis between the linear and non-linear method in determining the kinetic parameters was conducted.

Methods

1.1. Materials

All chemicals - $Mg(NO_3)_2{\bullet}6H2O$, $Al(NO_3)_3{\bullet}\ 9H_2O$, NaOH, and HCl were purchased from Merck Co. Inc., Germany, $NH_3{\bullet}H_2O$ and p-hydroxybenzoic acid were purchased from Sigma-Aldrich and nitrogen sufficiently high purity (99,99%) was purchased from CV Perkasa (Indonesia) - are of analytical grade and used as received without further purification. $[AuCl_4]^-$ solution ($HAuCl_4$) was prepared by dissolving of gold (99.99%) in aqua regia solution. Gold (99.99%) were purchase from PT. ANTAM Indonesia. The *p*-hydroxybenzoate intercalated Mg/Al-hydrotalcitewas prepared by co-assembly process [6]. Absorption spectra were obtained using aAtomic Absorption Spectroscopy.

1.2. Batch equilibrium and kinetic studies

Into a series of flasks containing 10 mL of $[AuCl_4]^-$ 100 mg/L solutions at optimum pH obtained from the above experiment, 10 mg of *p*-hydroxybenzoate intercalated Mg/Al-hydrotalcite was added. The flasks were then subjected to continuous stirring at contact times 0, 10, 20, 30, 60, 90, 120, 150, 200, 250, 300 and 350 min. At each selected contact time, a sample from the respective flask was immediately taken, filtered through a Whatman 42 filter and the remaining concentration of $[AuCl_4]^-$ in the filtrate was analyzed by AAS.

Results and Discussions

The ability of p-hydroxybenzoate intercalated Mg/Al-hydrotalciteto remove $[AuCl_4]^-$ was carried out at temperature 25 ^{0}C and initial medium acidity equivalent to pH 3. To describe adsorption data obtained under non-equilibrium conditions, two kinetic models, namely the pseudo-first-order (PFO) and the pseudo-second-order (PSO) were used [12-17]. The kinetic rate equations following PFO can be representated as

$$\frac{dq_t}{dt} = k_1(q_e - q_t) \quad (1)$$

Where k_1 is the rate constant.

Integrating Eq. (1) for boundary conditions $t = 0$ and $t = t$ and $q_t = 0$ and $q_t = q_t$, gives [16]:

$$\log(q_e - q_t) = \log(q_e) - \frac{k_1}{2{,}303}t \quad (2)$$

The constants can be determined experimentally by plotting of $\log(q_e - q_t)$ against t.

Likewise, the reductive adsorption of $[AuCl_4]^-$ following PSO kinetics can be represented as

$$\frac{dq_t}{dt} = K_2(q_e - q_t)^2 \quad (3)$$

Where k_2 is the rate constant.

Integrating Eq. (3) for boundary conditions $t = 0$ and $t = t$ and $q_t = 0$ and $q_t = q_t$, gives [17]:

$$q_t = \frac{k_2 q_e^2 t}{1 + k_2 q_e t} \qquad (4)$$

Eq. (4) can be linearized to at least five different linear forms as shown in Table 1. Table 1 also shows the non-linear form of Eq.

Table 1.*Linear and non-linear of pseudo-first-order and pseudo-second-order equations*

Kinetic model	Non-linear	Linear	Type
Pseudo-first-order	$q_t = q_e(1 - e^{-k_1 t})$	$\log(q_e - q_t) = \log(q_e) - \frac{k_1}{2,303} t$	PFO
Pseudo-second-order	$q_t = \frac{k_2 q_e^2 t}{1 + K_2 q_e t}$	$\frac{1}{q_e - q_t} = \frac{1}{q_e} + k_2 t$	PSO-1
		$\frac{t}{q_t} = \frac{1}{k_2 q_e^2} + \frac{1}{q_e} t$	PSO-2
		$\frac{1}{q_t} = \frac{1}{q_e} + \frac{1}{k_2 q_e^2} \frac{1}{t}$	PSO-3
		$q_t = q_e - \frac{1}{k_2 q_e} \frac{q_t}{t}$	PSO-4
		$\frac{q_t}{t} = k_2 q_e^2 - k_2 q_e q_t$	PSO-5

Data analysis from experimental $[AuCl_4]^-$ removal by p-hydroxybenzoate intercalated Mg/Al-hydrotalciteaims to determine adsorption kinetics of $[AuCl4]^-$ removal process. The analysis was performed by matching $[AuCl_4]^-$ data removal to PFO and PSO kinetics models. Data matching was carried out through linear and non-linear models. Linear model matching was done by determining the R^2coefficient which can be obtained automatically from Microsoft® Excel softwarebyplotting according to the Table 1. While non-linear model matching was done using the solver add-in feature available in Microsoft® Excel software. Determination of R^2coefficient fornon-linear model was done using Eq. (5).

$$R^2 = \frac{\Sigma(qt_{cal} - \overline{qt_{exp}})^2}{\Sigma(qt_{cal} - \overline{qt_{exp}})^2 + \Sigma(qt_{cal} - qt_{exp})^2}$$

qt_{exp} and qt_{cal} are amount of adsorbate which was adsorbed by the experimental and the adsorbatewhich was adsorbed by the calculation, respectively.

Study of the removal kinetics of $[AuCl_4]^-$ by linear model was studied by plotting the graph as shown in Fig. 1, while non-linear model of the two kinetics is presented in Fig. 2.

IOP Conf. Series: Materials Science and Engineering **990** (2020) 012017 doi:10.1088/1757-899X/990/1/012017

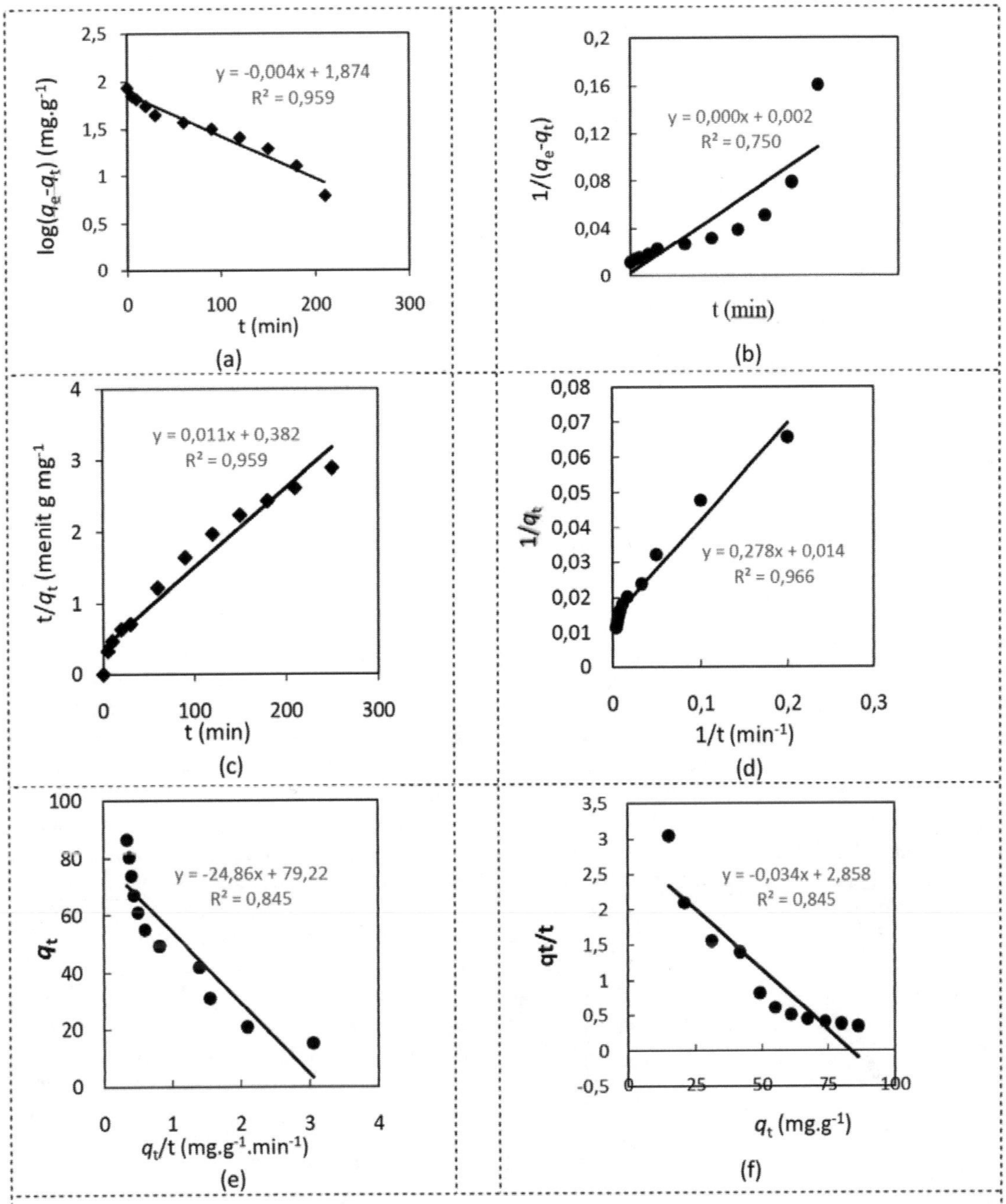

Figure 1. Experimental data and the fitted linear (a) PFO, (b) PSO-1, (c) PSO-2, (d) PSO-3, (e) PSO-4, and (f) PSO-5

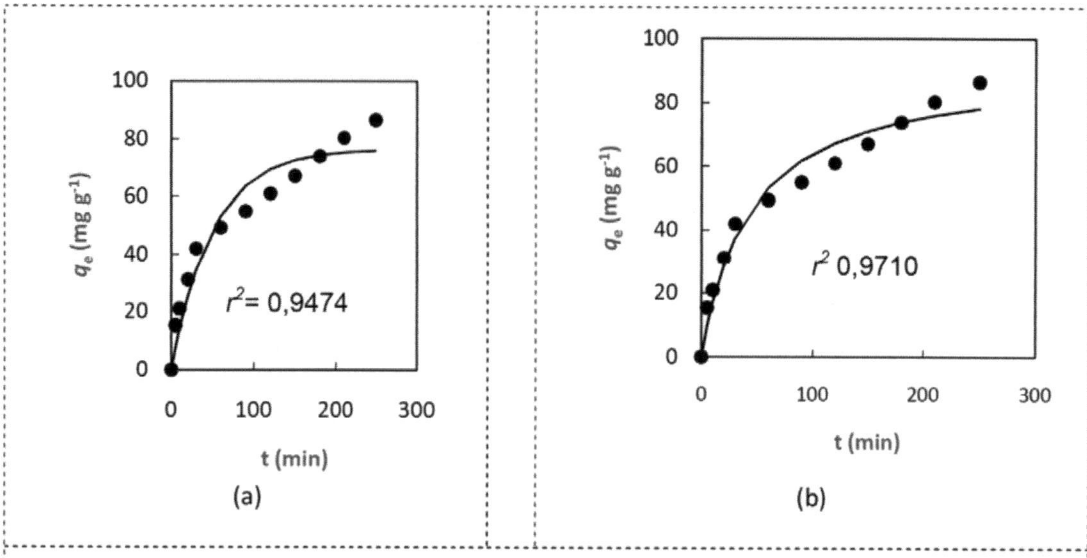

Figure 2. Experimental data and the fitted non-linear (a) PFO and (b) PSO

Fig. 1 shows that matching result of $[AuCl_4]^-$ removal by *p*-hydroxybenzoate intercalated Mg/Al-hydrotalcite using linear model produces a double result where the R^2coefficient of the PFO kinetics was greater than R^2coefficient of PSO-1, PSO-4, and PSO-5, but smaller when compared to R^2coefficient of PSO-2 and PSO-3. This can lead to errors in drawing conclusions, whether following PFO or PSO kinetics. The difference in matching results depends on how the kinetics equation was transformed into a linear form, which would have an impact on changing the error distribution so that the model parameter deviates [18]. In addition, the linear method which was based on theassumption that the scatter vertical points around the line were following the Gaussian distribution and the error distribution of each value on the X axis had a uniform value [19]. This rarely occurred or even practically could be said to be impossible in kinetics, as it was known that all adsorption kinetics models are non-linear. The linear method emphasized the distribution of errors only along the y-axis, each of the types of a kinetic model would produce different parameter values even with the same experimental data. Therefore, the linear method was not appropriate in predicting the suitability of a kinetics model for experimental data specifically and was unable to provide a good basic understanding of kinetics in the adsorption system so that it can produce wrong conclusions.

Table 2.*Kinetic parameters obtained by using the linear and non-linear methods.*

Kinetic model	*Eq.*	*k_1x 10^{-2} (min^{-1})*	*k_2x 10^{-4} (g.mg^{-1}.min^{-1})*	*R^2*
Linear pseudo-first-order				
Linear PFO	*5*	*1.0364*		*0.9593*
Linear pseudo-second-order				
Linear PSO-1	*6*		*50*	*0.7505*
Linear PSO-2	*7*		*3.2778*	*0.9596*
Linear PSO-3	*8*		*7.0453*	*0.9662*

Linear PSO-4	*9*		*5.0772*	*0.8459*
Linear PSO-5	*10*		*4.7574*	*0.8459*
Linear pseudo-first-order				
Non-linear PFO	*11*	*1.9871*		*0.9474*
Non-linear pseudo-second-order				
Non-linear PSO	*12*		*2.4938*	*0.9710*

In non-linear form, Fig. 2 and Table 2 show that PSO kinetic model with R^2coefficient value of 0.9710 gets a more appropriate result than the PFO kinetic model which has an R^2coefficient value of 0.9474. The use of non-linear models avoids the transformation process into linear form so that each experimental data has the same error structure. The matching results show that the removal rate of $[AuCl_4]^-$ by *p*-hydroxybenzoate intercalated Mg/Al-hydrotalcite follows the pseudo-second-order kinetics model, where $[AuCl_4]^-$ removal rate constant was 2.4938 x 10^{-4} g.mg^{-1}.minute^{-1}.

4. Conclusion

The removal rate of $[AuCl_4]^-$ by p-hydroxybenzoate intercalated Mg/Al-hydrotalcite was found to be well represented by the pseudo-first-order and pseudo-second-order kinetic. The matching results show that the removal rate of $[AuCl4]^-$ by p-hydroxybenzoate intercalated Mg/Al-hydrotalcite follows the pseudo-second-order kinetics model, where $[AuCl_4]^-$ removal rate constant was 2.4938 x 10^{-4} g.mg^{-1}.min^{-1}. This study shows that the bestfitting non-linear forms of the pseudo-first-order and pseudo-second-order kinetic models were superior to the linear forms. The model parameters may be distorted when the non-linear equations were transformed to linear forms. Therefore, the non-linear method should be primarily adopted to obtain the adsorption parameters.

Acknowlegment

The financially support from the Ministry of Finance, Republic Indonesia, through the Indonesia Endowment Fund for Education (LPDP).

Referencess

[1] HoY.S., McKay, Can. 1998. *J. Chem. Eng.***76** 822–826.
[2] Kumar K. V, Ramamurthi V, Sivanesan S. 2005*J. Colloid Interface Sci*. **284**14–21.
[3] Vasanth K. K.. 2006. *Journal of Hazardous Materials*. **B137**1538-1544
[4] Manju G. N, Gigi M. C, and Anirudhan T. S. 1999*Ind. J. Chem. Technol*. **6**34–141
[5] Anirudhan T. S, and Suchithra P. S*Appl. Clay. Sci*.**42**, 214–223.
[6] Lutfi A. H, Sri J. S., and Bambang R. 2020. *Indo. J. Fund. Appl. Chem*. **5** 1-10
[7] Ho Y S. 2004. *Carbon*, 2004, **42** 2115–2116
[8] Vasanth K K, Sivanesan S. 2005. *J. Hazard. Mater*.**B123** 288–292
[9] Tsai S C, Juang K W. 2000. *Journal of Radioanalytical and Nuclear Chemistry*. **243**741–746
[10] Azizian S, Haerifar M, Basiri-Parsa J. 2007. *Chemosphere*. **68**2040–2046
[11] Lin J. and Wang L. Front. 2009. *Environ. Sci. Engin. China*. **3**320-324
[12] Wu. F. C, Tseng R. L, andJuang R. S. 2001. *Water Research*.**35** 613-618. 2001.
[13] Tan. I. A. W, Ahmad A. L, and Hameed B. H. 2008. *J. Hazard. Mater*.**154** 337-346. 2008.
[14] K. V. Kumar. 2006. *J. Hazard. Mater*.**137**1538–1544
[15] Pacławski K, and Sak T. 2015*J. Min. Metall., Sect. B Metall*., **51**133–142. 2015
[16] Lagergren S. 2000. *Handlingar*.**24**1–39

[17] HoY. S and McKay G. 2000. *Water Res.***34** 735–742
[18] Lin, J. dan Wang, L. 2009*Front. Environ. Sci. Engin. China*, **3** 320-324.
[19] Chowdhury, S. dan Das Saha, P., 2011, *Bioremed. J*, **15** 181–188.

Physico-mechanical Properties of Low-density GeopolymerMortarSynthesized Using Inexpensive Foam Agent

A Amri[1*], M Reski[1], Z Helwani[1], Aman[1], Y B Hendri[1],H Maizir[2], S Saputro[3], and J Wahyuadi[4]

[1]Department of Chemical Engineering, University of Riau, Pekanbaru 28293, Indonesia,
[2]Department of Civil Engineering, SekolahTinggiTeknologiPekanbaru, Riau, Indonesia,
[3]Department of Natural Science Education, Universitas SebelasMaret, Surakarta, Indonesia,
[4]Department of Metallurgy and Material Engineering, University of Indonesia, Depok, Indonesia.

E-mail: amun.amri@eng.unri.ac.id

Abstract. Low-density geopolymer(LDG) mortars have been synthesized through the use of an inexpensive car-wash foaming agent. LDG mortarswere made by mixing coal fly ash and sand with activator solution of NaOH and Na_2SiO_3. The slurry mixture was then mixed with a thick foam and stirred using a hand mixer. During the stirring process, the ordinary Portland cement (OPC)was added to the mixture as additive.Once homogeneous, the slurry mixture was then poured into mortar molds and left overnight at room temperature before heat treatment at 60 °C for 24 h. The geopolymer mortars formed were then characterized. The density measurement showed that the density of geopolymerdecreased to about half from the original density at the OPC compositions of 5-15% and the activator solution compositions of 20-26%. In this composition range, the highest compressive strength (2.15 MPa) was shown by the LDG synthesized using a 23% activator solution and 15% OPC.Further addition of activator solution decreased the compressive strength due to the increase of porosity. The thermal durability test indicated that the LDG structure was relatively stable up to a temperature of 300 C. A further increase in temperature would cause structural degradation.

1. Introduction

Research on low-density geopolymers(LDG) or lightweight geopolymers has attractedthe attention of researchers [1, 2]. It is driven by various needs, including the desire to expand its applications, such as acoustic panels, thermal insulation, building wall materials, etc.Alvarezayusoetal. and Chengetal. in [2] reported that the low-density geopolymers had several characteristics including excellent mechanical properties, good thermal resistance, good adsorption ability of toxic metals, as well as high acid resistance and fire resistance. Apart from these advantages, the raw materials are easily obtained and cheap from an electric power plant fly ash waste or other agro-industrial fly ash.

[1] amun.amri@eng.unri.ac.id

Content from this work may be used under the terms of the Creative Commons Attribution 3.0 licence. Any further distribution of this work must maintain attribution to the author(s) and the title of the work, journal citation and DOI.
Published under licence by IOP Publishing Ltd

The choice of manufacturing method will affect the quality of low-density geopolymer produced. There are two techniques in preparing of foam mortar, namely pre-foaming and mix-foaming techniques. In the pre-foaming technique, the foam is made separately from the mortar paste. The next step is mixing the paste and foam; meanwhile, the mix-foaming technique, from the beginning, the foam and pasta are mixed. Ducman and Korat[1] conducted a study on the manufacture of fly ash-based low-density geopolymers by comparing the addition of aluminum powder and hydrogen peroxide (H_2O_2) foaming agents. The results of this study indicated that the hydrogen peroxide (H_2O_2) had a better effect than aluminum powder, and the porosity test showed that the porosity of aluminum powder washigher than the hydrogen peroxide (H_2O_2) by 59% and 48%, respectively. Wu et al., [2] have created fly ash - metakaolin based ultra-lightweight foamed geopolymer (UFG) with the addition of H_2O_2 foaming agent. The results of this study indicated that the lightweight geopolymer showed better mechanical properties than the portland cement-based foam concrete at the same density. However, materials such as aluminum powder and hydrogen peroxide are quite expensive. Besides, hydrogen peroxide (H_2O_2) is quite dangerous for health if inhaled and exposed to the body and environment.

This work tries to create a low-density geopolymervia a pre-foaming technique using a car-wash foam agent that is cheap and safe for the environment. The effects of an increasing amount of activator solution and OPC additiveused were studied. The results showed that the density of the geopolymer decreased to about half of its original density at 5-15% of OPCand 20-26% activator solution compositions. The compositionsare acrucial factor in obtaining optimal density and compressive strength values

2. Experimental

2.1 Geopolymers preparation

The activator solution was made by mixing 10 M NaOHand Na_2SiO_3 solutions with a ratio of Na_2SiO_3: NaOHof 2.5.The car-wash foam was prepared through the pre-foaming technique by mixing the car-wash foaming agent with water in a ratio of 1:30 and then fed into foam generator equipment. Solid fly ash and sand were mixed with activator solution with a solid to liquid ratio of around 4:1. This mixture was then poured into a container that already contained thick car-wash foam. Stirring was done quickly using a hand mixer while adding the OPC little by little to the dough until homogeneous. This dough was then poured into molds (10x10x10 cm^3) and left overnight. The geopolymer mortar formed was then heated in an oven at 60 C for 24 hours and then cooled. The amount of activator solution and the amount of OPC used were varied and repeated three times (by triplo)

2.2. Characterization

The density of the formed geopolymer wasdetermined by dividing the dry mass by volume. The compressive strength of geopolymers was tested using a compressive testing machine following ASTM C579-01 (2012). The thermal durability test was performed by monitoring of compressive strength of geopolymer after heating to high temperatures at 300 °C, 500 °C, and 700 °C in a furnace for 1 hour 30 minutes.

3. Results and Discussions

3.1 Density of Geopolymer

Table 1and Figure 1 show the densities of geopolymers produced by various percentages of activator solution used and OPC added. It can be seen thatthe geopolymers have a reasonably low density compared to the control sample by about half. The decrease of density is due to the high porosity in the geopolymers matrix created by the foam integrated into the matrix at the molding process. The formation of pores initiated by the foam is then maintained by OPC, which has a fast setting time of hardening. This pores formation is difficult to be reached without the addition of OPC due to the slow geopolymer setting time.

From Figure 1, it can also be seen that the density of geopolymers increases as the increase of OPC content in all variations of activator solution addition. It is understandable because OPC is the adhesive between the components in the matrix and, at the same time, solidifies them. Likewise, the densities of geopolymers also increase as the increase of activator solution up to 23%. It is due to the rise of NaOH and Na_2SiO_3 activator solution can disturb the stability of the foam bubbles, so they are easily broken into the liquid, which leads to a decrease in the number of pores and ends up with a higher density. However, when the activator solution increases to 26%, a strange phenomenon is observed, in which the density again decreases. So far, this is not fully understood. However, this is likely due to the high water content when the percentage of activator solution increases. In a specific composition, this water will counteract or offset the negative influence of NaOH and Na_2SiO_3 on the foam, so that it will lead to foam stability. Therefore, the amount of foam, activator solution and OPC used must be precise to obtain optimal strength of lightweight geopolymers.

Table1. Densities of the produced geopolymer at various percentages of OPCaddition and activator solution

Activator solution (%wt)	OPC (%wt)	Density (kg/m^3)
	5%	1.150
20%	10%	1.158
	15%	1.175
	5%	1.149
23%	10%	1.185
	15%	1.193
	5%	1.037
26%	10%	1.072
	15%	1.087
Geopolymers (without foam addition) as control		2.070

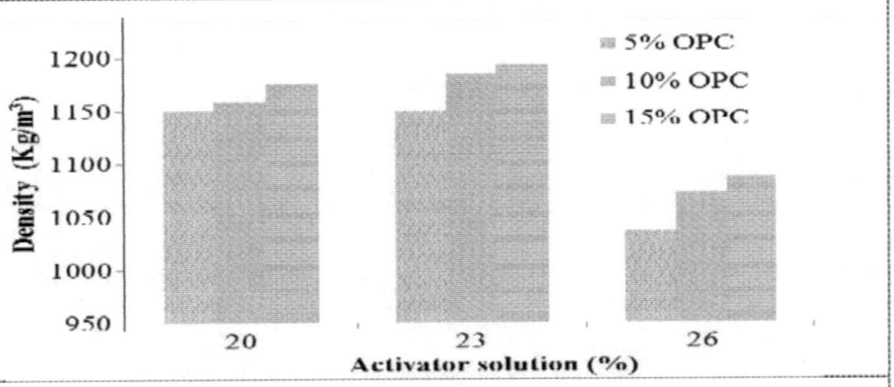

Figure 1. Densities of the produced geopolymers at various percentages of OPC addition and activator solution

3.2 Compressive strength

Figure 2 shows the compressive strength of low-densitygeopolymers(LDG) synthesized by various percentages of OPC addition and activator solution. It can be seen that the compressive strength of LDG increases with the increase of OPC addition for all activator solution compositions. It is well-understood that the more cement is added, the more compressive strength of the material will be[3].The highest compressive strength (2.15 MPa) is shown by LDG synthesized by OPC addition of 15% and activator solution of 23%. Linear to the densities data as discussed in the previous section, the compressive strength of LDG synthesized using activator solution of 26 % is lower compared to the LDG synthesized using activator solution of 20% and 23%. It is, of course, because their densities are the lowest than the others.

According to its nature and without being associated with the addition of foam or formation of pores, the addition of too much activator solution that passes through a certain composition can decrease the compressive strength of LDG[4].Phoo-ngernkham *et al.*[5] stated that the addition of excess activator solution could cause a decrease in the mechanical properties of geopolymers.It is similar to an increase in the ratio of water to cement. Special for LDG formation, the results of this study are also consistent with the work by Wu *et al.*[2]. They said that the addition of excess activator solution to the matrix will make the LDG tend to flow easily and thin, causing foam bubbles to move freely to form large cavities. These cavities can cause a decrease in mechanical properties of geopolymers, and when the curing process is applied, there is a shrinkage of LDG, resulting in cracks in the matrix, which will reduce the compressive strength of LDG. Based on the results, it can be concluded that the addition of OPC and activator solution at a certain maximum amount can increase the compressive strength of LDG.

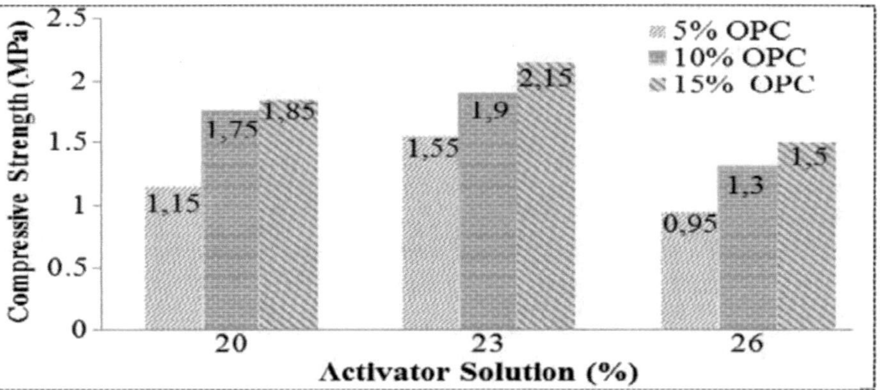

Figure 2. Compressive strength of low-densitygeopolymer (LDG) synthesized by various percentages of OPC addition and activator solution.

3.3 Thermal durability

To obtain a low-density geopolymer (LDG) that can be applied for various practical applications, besides testing the compressive strength at room temperature, it is also necessary to know the effect of heat treatment on the geopolymers compressive strength (thermal durability). Figure 3 shows the compressive strength of LDG,which is heat-treated for 1.5 hours at different temperatures.Geopolymers tested are geopolymers with a composition of 15% OPC and 23% activator solution, which gave the best compressive strength values, as discussed in Section 3.2. From the figure, it can be seen that the structure of the low-density geopolymeris relatively stable up to a temperature of 300 °C, which is indicated by a slight degradation in compressive strength of 11.62%. When heating exceeds a temperature of 300 °C, the low-density geopolymer begins to experience a significant degradation of strength, with a decrease in compressive strength of 35% at 500 °C and 72% at 700 °C. This decrease in compressive strength is due to the dehydroxylation reaction of the geopolymer. Where when applying high heat, the bonds to the geopolymer will be broken.

A visual inspection of the heat treatment results shows that the low-density geopolymer cracked only after heating above 500 °C. It is consistent with what was reported by Syamsidar et al. in [6], which said that the geopolymer has excellent heat resistance up to a temperature of 600 C. They also noted that the use of fly ash with high SiO_2 content and silica in the activator fluid (Na_2SiO_3) could provide excellent stability in maintaining the geopolymer structure. These results indicate that the low-density geopolymer has excellent resistance to high temperatures. Besides, the addition of OPC has a positive effect on durability [3]. Overall, based on the results in this work, it can be concluded that the produced low-density geopolymer can be used for conventional applications such as building wall materials or other low-temperature applications.

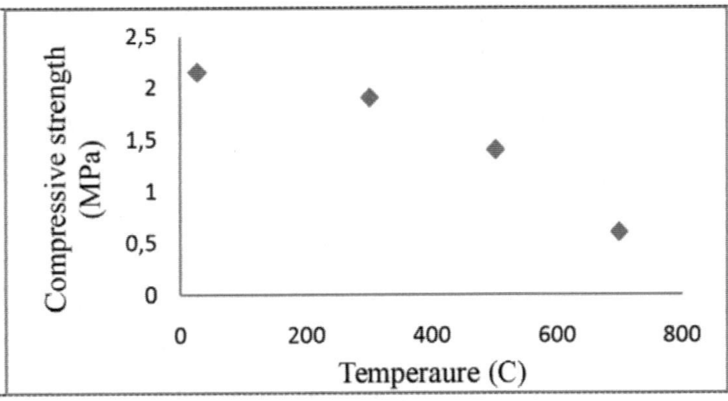

Figure 3. Compressive strength of lightweight geopolymer after heat treatment for 1.5 h at different temperatures

4. Conclusions

Low-density geopolymer (LDG) mortars have been synthesized by adding car-wash foam agent and adjusting the composition of ordinary Portland cement (OPC) as well as activator solutions of NaOH and Na2SiO3. The density of the LDG decreased to half of the initial density at the OPC composition between 5-15% and 20-26% activator solution. In this composition range, the highest compressive strength (2.15 MPa) was shown by the geopolymer synthesized using a 23% activator solution and 15% OPC additions. Further addition of activator solution decreased the compressive strength due to increased porosity. The thermal durability test indicated that the geopolymer structure was relatively stable up to a temperature of 300 C. A further increase in temperature would cause the structural degradation of the geopolymer.

Acknowledgment

Thanks to DIKTI for funding this research via KRUPT research grant 2019.

5. References

[1] Ducman Vand Korat L 2016 Materials Characterization **113**207-213.
[2] Wu J, Zhang Z, Zhang Y and LiD 2018 Construction and Building Materials**168**771-779.
[3] Mehta A and Siddique R 2017 Construction and Building Materials**150**. 792-807.
[4] Thomas M, Jewell R and Jones R 2017 Coal fly ash as a pozzolan, in Coal Combustion Products (CCP's)Elsevier121-154.
[5] Phoo-Ngernkham T, Phianghimai C, Damrongwiriyanupap N, Hanjitsuwan S, Thumrongvut J and Chindaprasirt P 2018 Advances in Materials Science and Engineering **2018**1-13.
[6]Syamsidar D, Amalia N, Nurfadilla and Subaer 2015 *Prosiding Pertemuan Ilmiah XXIXHFI* Jateng & DIY, Yogyakarta 49-50.

IOP Conf. Series: Materials Science and Engineering **990** (2020) 012019 doi:10.1088/1757-899X/990/1/012019

Drinking Water Production from Rainwater Using Radio Frequency Plasma System

R Desmiarti[1], E Sari[1], R R Vallepi[1], F S Wahyeni[1], M Y Rosadi[2] and A Hazmi[3]

[1]Department of Chemical Engineering, Universitas Bung Hatta, Padang, 25173 Indonesia

[2]Department of Engineering Science, Gifu University, Gifu, 501-1193 Japan

[3] Department of Electrical Engineering, Andalas University, Padang, 20362, Indonesia

E-mail: renitk@bunghatta.ac.id

Abstract. Indonesia has a large amount of rainfall and can be used as raw water of drinking water. A Radio frequencyplasma system radiation can produce active compounds (•OH^-, •O, •H^+, H_2O_2,O_3 etc) in water,the active compounds have a high oxidation potential and can kill microorganisms present in water (fecal coliform,total coliform and *Salmonella*). Plasma system is one way to produce drinking water in terms of health aspects because in the process the tool does not contact with the material directly, so the possibility of contamination is small. The purpose of this study is to remove microorganisms in rainwater using plasma radio frequency system continuously. Samples will be filtered using pure and mix polypropylene cartridge filter with a pore size of 1 μm and then contact in plasma system. Plasma is generated by applying a frequency of 0,16 MHz through a glass reactor with a thickness of 2 mm which is wrapped by a 1 mm copper wire. The results show that the removal microorganism in rainwater using plasma coupled with filtration using pure polypropylene filter reached 100% for total coliform, fecal coliform and *Salmonella*.While the removal microorganisms in rainwater using plasma coupled with filtration using mix polypropylene reached 70–100%, 85–100% and 80–100%, for total coliform, fecal coliform and *Salmonella*, respectively.

1. Introduction

Water is a primary need for all living things. Water is used to meet human needs, including for drinking, cooking, and washing. If the water needs have not met the drinking water standard,then it can have a major impact on health and social. Currently, a lot of drinking water treatment is used to get healthy and safety of drinking water. The water has certain standard requirements such as physical, chemical, bacteriological and radiological requirements set by the Indonesian Minister of Health No. 492/Menkes/Per/IV/2018. The raw material of drinking water from river also have natural estrogens [1,2]. The drinking water treatment process can be taken from the nearest water source, which is well water. Well water

[1] renitk@bunghatta.ac.id.

Content from this work may be used under the terms of the Creative Commons Attribution 3.0 licence. Any further distribution of this work must maintain attribution to the author(s) and the title of the work, journal citation and DOI.
Published under licence by IOP Publishing Ltd

has the potential to be processed and made into drinking water in accordance with the provisions. Well water is also used as raw water for refill drinking water depots. Depot drinking water treatment that uses the ultraviolet (UV) system and the use of filters in filter tubes, these tubes consist of sand filter media and activated carbon. In line with the development of water treatment technology, some refill drinking water depots use a reverse osmosis system to process water. However, the processing of UV light has many disadvantages such as the addition of chlorine or ozone after the UV process, it is relatively difficult to determine the UV dose, the formation of biofilms on the surface of the lamp and there is still the potential for photoreactivation in pathogenic microbes that have been processed with UV. Reverse osmosis has disadvantages such as eliminating some minerals that are useful in water, requires high energy because this process operates at a pressure of 10–60 bar. Therefore, water treatment with UV and RO does not guarantee producing drinking water with good quality in accordance with drinking water quality requirements.

The latest research from Padang City Health Departmentprovides that 666 drinking water depots were found to be 18% ineligible and suitable for consumption [3]. The excessive use of ground water can cause a decrease in the surface of the land, so that it requires a quick and appropriate solution. One solution is to find alternative water sources that can substitute for ground water functions, the most potential alternative water source in Indonesia is rainwater. Indonesia received rainfall of 2,000-4,000 mm/year and varies greatly in various regions. In Padang, the average monthly rainfall is 405.6 mm/month with an average of 17 days of rain per month [4]. With high rainfall intensity, rainwater could be transformed into clean water, therewith drinking water as an alternative water source.

McMichael *et al.* [5]used photo electrochemical reactor (PEC) with a compound parabolic collector (CPC) to test the electrochemically assisted photocatalytic (EAP) disinfection of rainwater under real sun condition. The targeted environmental strains of *Escherichia coli* and *Pseudomonas aeruginosa* showed the reduction of 5.5-log10 and 5.8-log10 for *E. coli* and *P. aeruginosa* with relatively low UV irradiance. Du *et al.*[6]harvested rainwater and filtered through gravity-driven membrane (GDM) with the permeate flux of 4.0 $L/(m^2\ h)$ and showed the decreased of bacterial abundance within the permeate $((8.45 \pm 0.11)x10^2$ cells/mL) and also could produce a permeate that was almost free of particles [6]. Filtration technique using a metal membrane was designed and developed for efficient and safe use of rainwater. The study showed the high treatment efficiency of microorganism and particulates with the combination of ozone bubbling as aeration which considered to reduce membrane fouling and inactivate microorganisms [7]. Biosand filters have been demonstrated to inactivate harmful microorganisms as well as UV irradiation that has shown over 99.9% inactivation efficiency for *Cryptosporidium parvum* oocysts and *Giardia lamlia* cysts at low UV dose [8].

Numerous alternative disinfection has been suggested to purify rainwater from microorganisms. Radio frequency plasma system is an advance treatment used to produce drinking water with a small possibility of contamination. Plasma can generate oxidizing species radical(•OH^-, •O, •H^+)and molecules (O_3 and H_2O_2) [9]. Theseoxidizing species have a high oxidation potential to disintegrate bacterial cells and decompose organic compounds in water [10]. Also, plasma can produce ultraviolet light and shock waves which can also decompose organic compounds [8].

Previous studies have been conducted with the investigation of the removal of pathogenic microorganisms using plasma system. However, the removal of microorganisms in water still below 100% [11, 12, 15]. Therefore, this study is objected to investigate the application of radio frequency plasma coupled with filtration process to produce drinking water from rainwater based on the evaluation of microorganisms' removal efficiency using different material of cartridge filter.

2. Materials and methods

2.1 Rainwater sample

The rainwater was collected from the roof catchment in two different sampling locations located in Padang, West Sumatra, Indonesia. A rainwater was collected from Sinar Melayu Residence located in Jalan Gajah Mada Dalam, Padang. B rainwater was collected from Badan Pengawas Keuangan dan Pembangungan (BPKP) Residence 2A Nanggalo located in Jalan Shinta Kenanga, Padang, Indonesia. The distance between two site is 2.5 Km.

2.2 Cartridge filter used in the experiment

The filters used in this experiment are shown in Table 1.

Table 1. Cartridge Filter Material Type

Type of Filter	Material and Excellence
Filter I	• Made from Mix polypropylene • To filter turbidity and large, small particles in water • Pore size 1 μm
Filter II	• Made from Pure polypropylene • To filter turbidity and large, small particles in water • Pore size 1 μm

2.3 Radio frequency plasma-filtration system

The collected rainwaters were transferred to the raw water tank and pumped with a flow rate of 100, 150 and 200 mL/min through cartridge filter and flowed through plasma reactor coupled with copper wire under the frequency of 1.5MHz and electric current of 3A. Two different cartridge filters used in this study were 1 μm pure polypropylene and mix polypropylene filter (Table 1). The illustration of the process is shown in Figure 1. The treated rainwaters defined as drinking water were collected in the drinking water tank and analysed for total coliform, fecal coliform and *Salmonella*.

The number of microorganisms was measured with the plate count method [13,14]. The method used is the same as that describe in Desmiarti et al [15]. To analyse the number of pathogen bacteria, 1 mL treated water was poured onto a petri dish with a diameter of 5 cm, containing 4 mL of melted nutrient agar and the kept at 37 in an incubator for 24 h. The total number of bacteria was counted by enumerating purple, pink and white colonies formed as a fecal coliforms, total coliforms and *Salmonella*, respectively, in a colony-forming unit (CFU/mL.)

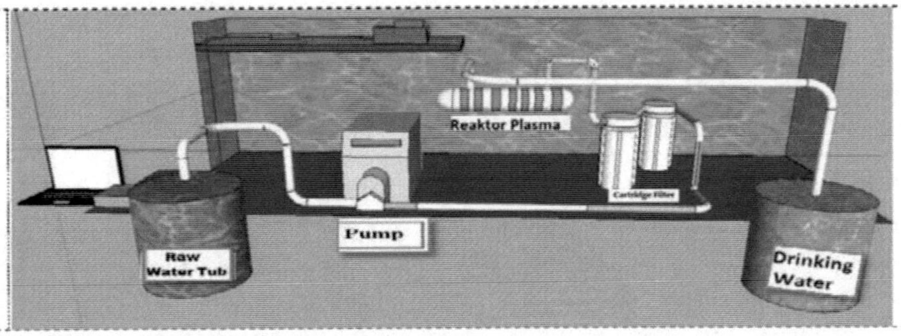

Figure 1. Experimental set-up of radio frequency plasma-filtration system

3. Results and discussion

3.1. Effectofflow rate on the removal of Salmonella *through a pure polypropylene filter*
A pure polypropylene filter could remove total coliform and fecal coliform in the rainwater, while *Salmonella* remained in the rainwater after filtration (Table 1). Figure 2 showed that the different flow rate caused a slight significant effect on removal of *Salmonella* in the first 10 min after plasma treatment. However, the removal efficiency of 100% was achieved after 40 min of treatment at all flow rate. The flow rate of 100 mL/min showed the highest removal rate in the first 10 min. This result is consistent with the previous study [14] that the removal of microorganisms will increase with the decrease of the flow rate during radio frequency plasma treatment system.

The efficiency on the removal of microorganisms also could be seen in the value of electric current and voltage which is determining the ionization process that occurs in water, the greater electric currentis applied, the ionization energy will release electrons is even greater. The voltage will also affect the amount of electric field that occurs in the reactor coil, as well as the electric current that affects the magnitude of the magnetic field[16].

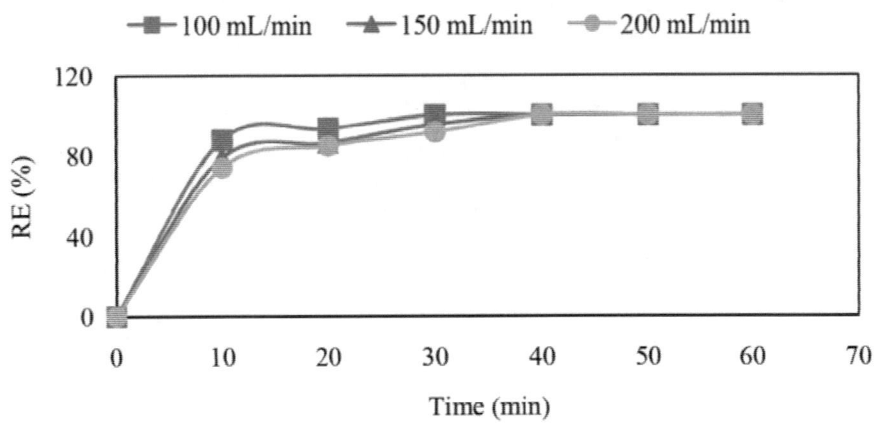

Figure 2. Effect of flow rate on the removal of *Salmonella* in rainwater after 60 minutes plasma treatment (RE: removal efficiency)

IOP Conf. Series: Materials Science and Engineering **990** (2020) 012019 doi:10.1088/1757-899X/990/1/012019

The combination of magnetic and electric fields is a factor that influences the electron ionization which serves to trigger the formation of plasma (oxidizing species) which kills microorganisms in water[17]. In this study, the picoscope was used with an electric current of 16.1 A and the voltage of 100 V. The high removal rate in after 10 min treatment at a flow rate of 100 mL/min could be supported finding by the influence of preferable electric field that formed during plasma treatment. The advantage of this combination of cartridge filter and radio frequency plasma compared to the DBD system is that the sample does not come in direct contact with the plasma and its application is simple. When comparedwith the electrolysis process which conducts in long treatment time, radio frequency plasma system requiresshort treatment time (60 minutes).

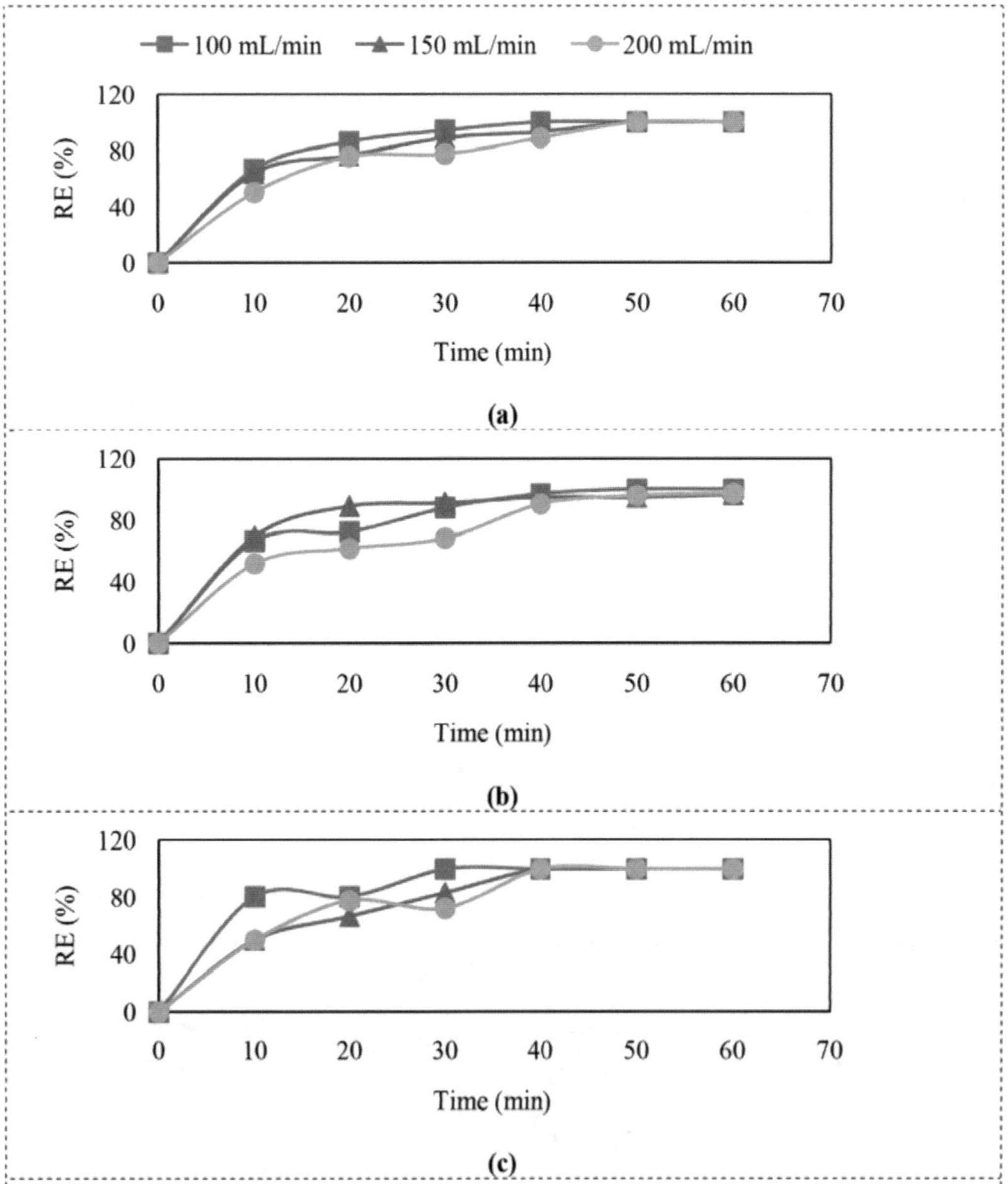

Figure 3. Effect of flow rate on the removal of (a) fecal coliform, (b) total coliform and (c) *Salmonella* after 60 minutes plasma treatment (RE: removal efficiency)

3.2. Effectof flow rate on the removal of fecal coliforms, total coliforms and Salmonella *through a mix polypropylene filter*

Figure 3 showed the comparison of the removal efficiency on different microorganism's consortium. Fecal coliform, total coliform and *Salmonella* has successfully 100% removed after 60 min with all different flow rate. This suggested that prolonged contact time could promote bacterial cells disintegration and organic compound decomposition by ultraviolet rays and shock waves produced during plasma treatment system in water [9, 12]. Additionally, an electric current of 16.1 A and voltage of 100 V applied in plasma system could also promote the rapid death rate (data not shown) of microorganisms accompanied with the formation of electric filed that formed active species during treatment which can disintegrate bacteria cells in the water [9].

The removal efficiency of 66.7 and 80.6% was achieved in the first 10 min at a flow rate of 100 mL/min for fecal coliform and *Salmonella*, respectively. This finding indicated that fecal coliform and *Salmonella* were susceptible to the electric wave produced during plasma treatment with low water flowrate.However, the removal efficiency offecal coliforms did not constantly achieve and there was a decrease in efficiency at some time during plasma contact in water. The inconstant removal of microorganisms was probably due to ununiform pores of mix polypropylene filter that caused microorganisms drifted during the filtration process.

3.3. Performance of combination treatment with radio frequency plasma and different filter type

Table 2 showed the combination treatment of radio frequency plasma system with different cartridge filter in the removal of microorganisms. Total coliform and fecal coliform were no longer present in the water because this pure polypropylene consists of fibrous material and uniform filter pores, making it more efficient for the water filtration system. This type of filter can filter total coliform and fecal coliform that are retained in the pores of the filter.Whereas*Salmonella* removal efficiency was 100% achieved in the different rainwater source at all different flow rates. Meanwhile, the combination treatment of radio frequency plasma system with mix polypropylene filter could achieve the removal efficiency of 96, 100 and 100% for total coliform, fecal coliform and *Salmonella*, respectively, at a flow rate of 100 mL/min in A rainwater. The removal efficiency of 100% was achieved for all microorganism at a flow rate of 100 mL/min in B rainwater. This finding indicated that microorganisms could be more susceptible against plasma and there was more contact with plasma to disintegrate bacteria cells with a low flow rate.

Table2. Removal of efficiency of microorganisms after 40 minutes of plasma treatment

Filter	Rainwater	Flowrate (mL/min)	Removal efficiency (%)		
			Total coliform	Fecal coliform	*Salmonella*
Pure Polypropylene	A	100	-	-	100
		150	-	-	100
		200	-	-	100
	B	100	-	-	100
		150	-	-	100
		200	-	-	100
Mix Polypropylene	A	100	96	100	100
		150	70	100	87.5
		200	70	89	83
	B	100	100	100	100
		150	94.6	93	100

(a)

	200	90	88.9	100

4. Conclusion
A combination treatment using radio frequency plasma and filtration process in different flow rate was investigated to remove fecal coliform, total coliform and *Salmonella* in rainwater. Different cartridge filters provided a significant effect on the presence of microorganisms, where fecal coliform and total coliform were undetected in the filtrate after filtration through pure polypropylene filter. The rapid removal efficiency was achieved using mix polypropylene filter with a flow rate of 100 mL/min in the first 10 min of treatment, indicating more microorganisms were contacted with oxidizing species at a low flow rate. A 100% removal efficiency for *Salmonella* was achieved in the 40 minutes of treatment, indicated that *Salmonella* was more susceptible against plasma compared to fecal coliform and total coliform. To assess a drinking water quality standard, further investigation is required, concerning the evaluation of organic and inorganic matter present in the rainwater.

References
[1] Desmiarti R, Li J, Li F 2012, Water, Air and Soil Pollution 223(6) 3307-3320.
[2] Desmiarti R, Li J, Li F-S 2011, ITB Journal of Engineering Sciences 43B(2) 153-160.
[3] Padang City Health Department 2016 *Departement Kesehatan Kota Padang.*
[4] Meteorological, Climatological and Geophysical Station 2018 *Badan Pusat Statistic Kota Padang (Statistics Padang)*
[5] McMichael S, Waso M, Reyneke B, Khan W, Byrne J A and Fernandez-Ibanez P 2021 *App. Cat. B: Environ.***281** 119485
[6] Du X, Xu J, Mo Z, Luo Y, Su J, Nie J, Wang Z, Liu L and Liang H 2019 *Sci. Tot. Env.***697** 134187
[7] Kim R H, Lee S and Kim J O 2005 *Desalination***177**(1–3) 121–132
[8] Gadgil A 1998*Ann. Rev. En. and the Environ.***23**(1) 235–286
[9] Chiemchaisr C, Passananon S, Ngo H H and Vigneswaran S 2009*Desalination***234** 335-343
[10] Wang X, Huang Q, Ding S, Liu W, Mei J, Hunag Y, Luo J, Lei L and He F 2020 *Separ. & Purif. Technol.***240** 116659
[11] Desmiarti R, Hazmi A, Trianda Y, Sari, E, 2015*Research Journal of Pharmaceutical, Biological and Chemical Sicences,* 6(1) 889-897.
[12] Desmiarti R, Hazmi A and Trianda Y 2015 *Modern App. Sci.* **9**(7) 80–85
[13] Hussain S N, Trzcinski A P,Asghar H M A, Sattar H, Brown N W and Roberts E P L 2016*J. Indust.& Eng. Chemist.***44** 216–225
[14] van den Akker B,Trinh T, Coleman H M, Stuentz R M and Le-Clech P2014 *Biores. Technol.***55** 432–437
[15] Desmiarti R,Hazmi A, Emeraldi P, Martynis M, Trianda Y and Sutopo U M 2018 *MATEC Web of Conf.***156** 03038.
[16] Lim S S, Fontmorin J, Izadi P, Daud W R W, Scott K and Yu E H 2020*Intern.J. Hydro. En.***45**(4) 2557–2568
[17] Rezaei F, Vanraes P,Nikiforov A, Morent R and de Geyter N 2019Applications of plasma-liquid systems: A review *Materials***12**(17) 2751

IOP Conf. Series: Materials Science and Engineering **990** (2020) 012020 doi:10.1088/1757-899X/990/1/012020

Development of Electrically Conductive Cellulose Nanofiber Film Composited with Carbon Nanotubes Using a Spray Method

M Tominaga[1], K Kuwahara,[1] M Tsushida,[2] K Shida[2]
[1]Graduate School of Science and Engineering, Saga University, Saga 840-8502, Japan
[2]Faculty of Engineering, Kumamoto University, Kumamoto 860-8555, Japan

E-mail: masato@cc.saga-u.ac.jp

Abstract. Nanocellulose is made from wood, cotton, cellulose-rich materials, and organisms by mechanical methods or chemical dissolution.Nanocellulosefilms are a flexible substrate with excellent thermal and chemical stability.These excellent properties are useful as electrodes for wearable type devices that replace plastic films. In addition, cellulose is both biodegradable and eco-friendly. Problem is that the cellulose film does not have electrical conductivity. In this study, a nanocelluloseconductive film made of composite carbon nanotubes was produced by a simple spray method.It was investigated whether this conductive film is useful as an electrode for electrochemical measurement.

1. Introduction

Cellulose is the most abundant carbohydrate on earth. Nanocellulose is made from wood, cotton, cellulose-rich materials, and organisms by mechanical methods or chemical dissolution [1-6].Nanocellulose can be roughly classified as cellulose nanofiber (CNF), cellulose nanocrystals, and bacterial nanocellulose. Nanocellulosefilms are a flexible substrate for preparation of electronic devices with excellent thermal and chemical stability.[1,5,6]. They are also biodegradable, eco-friendly, and readily disposable like paper [1,2].These properties are important for substrate materials used in wearable disposable devices. Because many types of flexible, wearable, and disposable electronic devices have been developed for use as chemical and physical sensors, and many solar cells contain plastics that are difficult to biodegrade [7-9]If the CNF-based thin film electrode can be made, a wearable, disposable, environment-friendly device electrode can be made. However, CNF does not have electrical conductivity. In this study, in order to fabricate an electrically conductive CNF, a composite electrode of CNF and carbon nanotubes (CNTs) was prepared by using simple spray method, and the electrochemical characteristics of the prepared electrode were examined.

2. Experimental

2.1 Materials

Nanoforest-S BB made from bamboo using the aqueous counter collision (ACC) method was used as the CNF, and was kindly supplied by Chuetsu Pulp and Paper Co., Ltd. (Tokyo, Japan). The ACC-

[1] masato@cc.saga-u.ac.jp.

Content from this work may be used under the terms of the Creative Commons Attribution 3.0 licence. Any further distribution of this work must maintain attribution to the author(s) and the title of the work, journal citation and DOI.
Published under licence by IOP Publishing Ltd

treated CNF has both surface properties of hydrophobic and hydrophilic[10,11].Multi-walled carbon nanotubes (MWCNTs, outer diameter 10 (± 1) nm, inner diameter 4.5 (±0.5) nm, and length 3–6 mm) was obtained from Sigma-Aldrich. Potassium hexacyanoferrate(III) ($K_3[Fe(CN)_6]$) was obtained from FUJIFILM Wako Pure Chemical Industries. Water was purified with a Millipore Milli-Q water system. All other chemical reagents were of analytical grade and were used without further purification.

2.2 Preparation of electrically conductive MWCNT/CNF film
An electrically conductive MWCNT/CNF film was prepared as follows. First, 6-47 mg CNF and 4-20 mg MWCNTs were dispersed in 10 mL of water using an ultrasonic homogenizer (Sonifier 450AA, Branson Ultrasonics Division of Emerson Japan Ltd). Next, 2.5 mL of the MWCNT/CNF dispersion solution was sprayed several times to an Al plate of 15×15 mm under heating at approximately 200 °C on a hotplate. Then, the prepared thin film was pilled off from the Al plate. The thickness of the electrically conductive MWCNT/CNF film was approximately 200μm.

2.3 Instrumentation
Scanning electron microscopy (SEM) imaging was performed using a JSM-7600F (JEOL, Japan). Each sample was coated with approximately 5 nm of osmium by sputtering before the SEM measurements. Field-emission transmission electron microscopy (FE-TEM) was performed using a Tecnai F-20 (Philips Electron Optics, Netherlands). Voltammetry measurements were performed with an electrochemical analyzer (ALS/CHI model 600A) with a conventional three-electrode cell. The reference electrode was Ag/AgCl/saturated KCl (+199 mV vs. the normal hydrogen electrode) and a platinum plate was used as the counter electrode. All potentials are reported with respect to Ag/AgCl/saturated KCl at 25 °C. Prior to the voltammetric measurement, UV-ozone treatment was performed to the prepared film surface using a UV–ozone system (OC-2503, Eye Graphics Co., Japan). Cyclic voltammograms were simulated with cyclic voltammetry simulation software (DigiSim 2.0, Bioanalytical Systems, Tokyo, Japan) [12].

3. Results and Discussion

3.1 Mixed dispersion solution of MWCNT and CNF
The mixed dispersion solution of MWCNTs and CNF did not show any separation or sedimentation after more than 2 weeks. Fig. 1 shows TEM images of the dispersed sample on a TEM grid. In these images, the CNF is dispersed in clusters measuring 10–50 nm in diameter, which is in good agreement with a previous report [11]. Interestingly, the dispersed MWCNTs are included into the CNF clusters as shown in Fig. 1b. The MWCNT surface was covered with the CNF. Formation of this supramolecular structure leads to good and stable dispersion of the MWCNTs in the aqueous solution.

The CNF used is prepared by using aqueous counter collision (ACC) method without any chemical modification [10,11]. The ACC-method treated CNF surface has both properties of hydrophobic and hydrophilic. For example, oil-in-water Pickering emulsions with long-term stabilities were easily prepared by just ultrasonically mixing aqueous ACC-prepared CNF with non-polar solvents [11].This result is due to the CNF surfaces of hydrophobic and hydrophilic. In this study, the MWCNT also was rapped with the hydrophobic surface of the CNF.

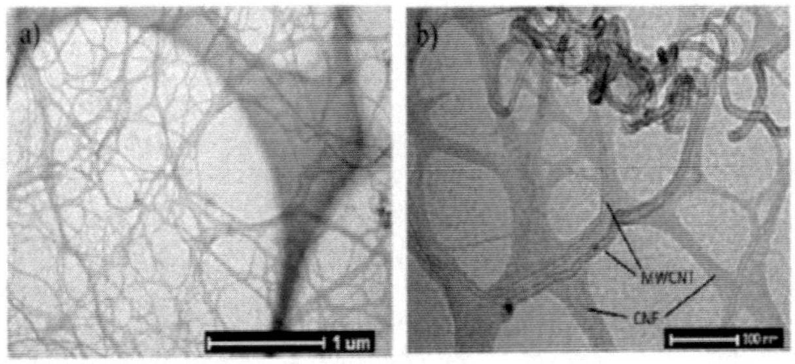

Figure 1. TEM images of the dispersed MWCNT/CNF sample. Wide range(low magnification) (a) and narrow range (high magnification) images(b).

3.2 Film structure and electrically conductivity

Fig. 2 shows the photographs of the MWCNT/CNF film, when the mixed solution of 10 mg MWCNT and 27 mg CNF (Fig. 2a), or the mixed solution of 10 mg MWCNT and 10 mg CNF (Fig. 2b) was used. The film of Fig. 2b was more black color than the film of Fig. 2a because of 2.7 times higher ration of MWCNT/CNF comparison to the film of Fig. 2a, which let us expected to be higher electrically conductivity of the film in Fig. 2b. The electrically conductivity was summarized in Table 1. Interestingly, the electrically conductivity was not strongly depended on the MWCNT/CNF ratio. The highest conductivity was observed the film prepared from 3 mg MWCNT and 7 mg CNF. This result let us suggest that thedispersed condition of the mixed solution is important. When the higher concentration solution of CNF and MWCNT, in other words, high ration of MWCNT/water and CNF/water, was used for the spraying solution, the CNF and MWCNT were not dispersed well each other in a micro revel, even if it appeared to be well mixed by eye. It is considered that the network of conductive MWCNTs is not well posted when the solution that is not dispersed well is used.

Fig. 3 shows the morphology of the surface and a cross section of the MWCNT/CNF film (200 Ωcm^{-1}), when the mixed solution of 3 mg MWCNT and 7 mg CNF was used. The surface morphology was similar to carpet fiber. Random fibers with diameters of a few dozen nanometers were observed. The cross-section image (Fig. 3b) showed the structure was formed of many layers with nanometer-scale thickness. This laminated structure could have high conductivity in the horizontal dimension.This structure would be due to multi spraying and layering until the film was completed.

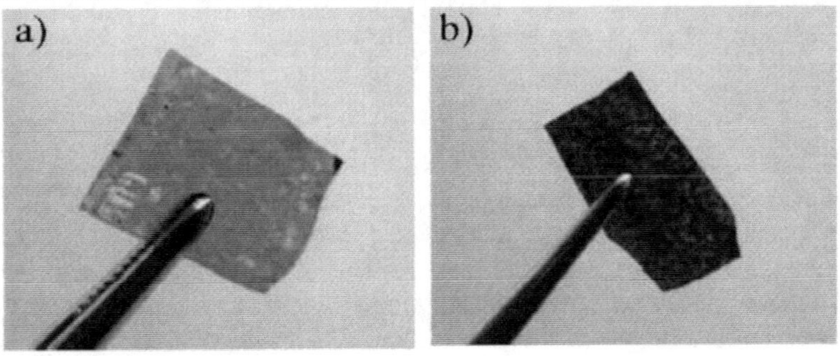

Figure 2. Photographs of the prepared MWCNT/CNF film. The (MWCNT:CNF) composition ratios of (a) and (b) were (10:27) and (10:10), respectively.

Table 1 Mixing ratio of MWCNT, CNF and water used in the spray method and electrical resistance of the produced film

MWCNT (mg)	CNF (mg)	Water (ml)	Resistance (Ω/cm)
4	6	10	1700
3	7	10	200
10	10	20	18000
10	20	10	1900
10	47	10	800

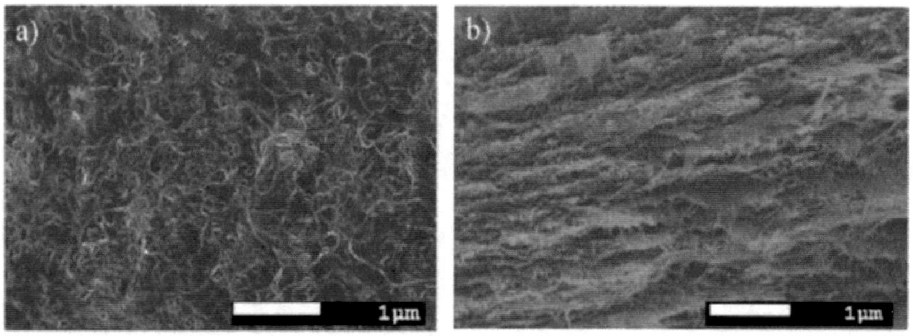

Figure 3. SEM images of the surface (a) and cross section (b) of the MWCNT/CNF film.

3.3.Electrochemical behaviors

To evaluate the performance of the MWCNT/CNF film (200 Ωcm^{-1}) as an electrode, cyclic voltammetry of $[Fe(CN)_6]^{3-}$ in 1 mol dm^{-3}KCl solution (pH 7) was conducted (Fig. 4). Separation between the anodic and cathodic peaks was approximately 300 mV. The redox reaction was a perfect diffusion-controlled process because the peak current was proportional to the square root of the potential sweep rate. To analyze the redox reaction, a voltammetric simulation was performed.[13] This gave a simulated voltammogram that was a good fit for the background (electrical double layer capacitance) subtracted voltammogram (Fig. 4). The redox potential ($E^{\circ\prime}$) and diffusion coefficient (D) were 0.27 (vs. Ag/AgCl/saturated KCl) and 1.4×10^{-5} cm^2 s^{-1}, respectively, and were in good agreement with previously reported values [14]. The heterogeneous electron transfer rate ($k^{\circ\prime}$) was estimated to be 6.5×10^{-5} cm s^{-1}, which was almost 1/20th that of the previously reported value [13]It was suggested that the electrode reaction of $[Fe(CN)_6]^{3-}$ was slightly inhibited by the CNF covered with the MWCNT. The estimated electrode surface area was 0.63 cm^2 even though the apparent electrode surface area was only 0.24 cm^2. The larger effective surface area could be attributed to the carpet fiber-like morphology of the surface. Even if the film was immersed in the solution for 1 week or more, it did not decompose and could be used sufficiently as an electrode.

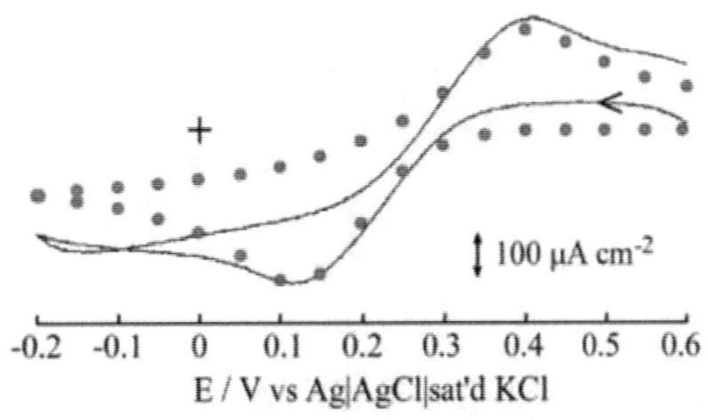

Figure 4. Background subtracted cyclic volta mmo grams of 1 mmol dm^{-3} $[Fe(CN)_6]^{3-}$ at the MWCNT/CNF film in 1 mol dm^{-3}KCl solution (pH 7) at a potential sweep rate of 20 mV s^{-1} and the simulated cyclic volta mmo gram (open circle) using the following parameters: active surface area, 0.4 cm^2; redox potential, $E°' = 0.27$; diffusion coefficient, $D = 1.4 \times 10^{-5}$ cm^2 s^{-1}; heterogeneous electron transfer rate, $k°' = 6.5 \times 10^{-5}$ cm s^{-1}; and transfer coefficient, $\alpha = 0.5$.

4. Conclusion

A CNF-based film electrode was manufactured using MWCNTs as an electrically conductive material by using a simple spray method.CNF prepared by the aqueous counter collision method gave good dispersion of the MWCNTs, which gave us the useful condition for film production by the spray method. When prepared by the spray method of a solution in which the mixing ratio and concentration of CNF and CNT were optimized, a conductive MWCNT/CNF film having a low electrical resistance could be prepared. It was also shown that the film does not decompose even when immersed in the solution for a long period of time and can be sufficiently used as an electrode in the solution.

Acknowledgment
We thank Chuetsu Pulp & Paper Co., Ltd. (Tokyo, Japan) for samples of Nanoforest-S BB.

References

[1] Mokhena T C and John M J 2020*Cellulose***27** 1149.
[2] Siro I and Plackett D 2010*Cellulose***17** 459.
[3] Sulaiman S, Mokhtar M N, Naim M N, Baharuddin A S and Sulaiman A 2015*Appl. Biochem. Biotechnol.***175** 1817.
[4] Rostami J, Mathew A P and Edlund U 2019*Molecules***24** article number 3147.
[5] Vilela C, Silvestre A J D, Figueiredo F M L and Freire C S R 2019 *J. Mater. Chem. A***7** 20045.
[6] Islam M S, Chen L, Sisler J and Tam K C 2018 *J. Mater. Chem. B* **26** 864.
[7] Choi S, Lee H, Ghaffari R, Hyeon T and Kim D-H 2016*Adv. Mater.* **28** 4203.
[8] Wen L, Li F and Cheng H-M 2016*Adv. Mater.***28** 4306.
[9] Khan S, Lorenzelli L, Ravinder S and Dahiya R S 2015*IEEE Sensors J.***15** 3164.
[10] KondoT,Kose R, Naito H and Kasai W 2014 *Carbohydrate Polymers***112** 284.
[11] Yokota S, Kamada K, Sugiyama A and Kondo T2019 *Carbohydrate Polymers* **226** 115293.
[12] Rodolph M, Reddy D P and Feldberg S W 1994*Anal. Chem.***66** 589A.
[13] Tominaga M, Hirata N and Taniguchi I 2005*Electrochem. Commun.* **7** 1423.
[14] Santucci R, Ferri T, Morpurgo L, Savini I and Avigliano L1998 *Biochem. J.***332**611.

IOP Conf. Series: Materials Science and Engineering **990** (2020) 012021 doi:10.1088/1757-899X/990/1/012021

Score Level Fusion Technique for Human Identification

M H Hamd[1] and Rabab A Rasool
Computer Engineering Dept., Al-Mustansiriyah University, Iraq

E-mail: dr.muthana@uomustansiriyah.edu.iq

Abstract. A multimodal biometric-system based score level fusion technique is proposed to construct a robust human identification system. Feature fusion can be implemented via different methods. In this paper, the score level fusion of face and iris traits are combined and re-classified at Equal Error Rate (EER) value to improve the individual unimodal systems performance for recognizing 80 subjects (40 subject per one face-iris dataset). The multimodal classification results are compared and evaluated comprehensively using four competitive feature extraction methods: Principle Component Analysis (PCA), Fourier Descriptors (FDs), Gray Level Co-occurrence Matrix (GLCM), and Local Binary Pattern (LBP). Also, a low-quality resolution of MMU1 iris database are considered in this work as an additional challenge for system robustness. The accuracy rate of GLCM and LBP methods satisfied 100% with ORL-CASIA-V1 combination datasets, while PCA and GLCM methods achieved 100% with the low-quality ORL-MMU-1 combination datasets, these results provide an evidence of how the multimodal biometric system could improve the overall unimodal systems performance. Also, the GLCM advances all other feature extraction methods by having the highest accuracy rate with ORL-CASIA-V1 and ORL-MMU-1 combined datasets.

1. Introduction

Multimodal biometric system is one of modest security system, it depends on concatenation more than one biometric trait in specific level to produce several advantages like, increasing the biometric security and improving the performance. Also, it solves system dependency on one biometric where each source of information is working separately and independently [1]. If any information source failed, the other one will not be affected [2]. The multi-biometric system can be performed with different techniques to support system performance, it may be a multi-instance, where several images are captured in different instances for the same trait and one sensor is used for performing this type, like captures several images for face in different facial expressions or iris images in different angles [3]. The multi-sensor type is utilized to capture images for same biometric trait [3], and the multi-algorithms which employsmulti-classifiers for same biometric trait, and the results from the different matching algorithms are combined to increase the system performance and reduces the dependency on one algorithm [3]. The two or more biometric traits are used for multi-modal human identification. For example, fusion of face and iris trait of same person to perform the identification. The advantage of each biometric system will improve the overall system advantages. Finally, the hybrid type is represented by combined of two or more multi-biometric systems in one system [4].

[1]dr.mumtana@uomustansiryah.edu.iq

Content from this work may be used under the terms of the Creative Commons Attribution 3.0 licence. Any further distribution of this work must maintain attribution to the author(s) and the title of the work, journal citation and DOI.
Published under licence by IOP Publishing Ltd

Muthana Hamd. and Marwa. M. in [5] proposed feature level fusion between face and iris traits using four feature extraction methods. The LBP, GLCM, FDs, and PCA method were applied for each trait separately to extract the features. The fusion was implemented using serial rule which produces new feature template. The classification stage was performed using ED. Also, three databases were utilized: CASIA-V1, MMU-1, and ORL. The suggested method achieved 100% accuracy rate using (LBP, GLCM) and 97.5% using (FDs, PCA) method. The developed face-iris multimodal biometric system in [6] was implemented by concatenating the two trait features in one vector. It utilized Gabor and LBP as a feature extraction method to form LGXP, also the average matching score was calculated using the distance measure and irrelevant pixels weightage. The experimental results provided better accuracy for multimodal recognition. The facial image information at higher and lower resolution were extracted and combined at different resolution for enhancement purpose [7]. The three fusion techniques: feature level, score level, and decision level were proposed by [8] to fuse the face and iris modalities using Log-Gabor method for feature extraction. Backtracking Search Algorithm was applied to improve the recognition rate by selecting the optimized weights for score level and feature level fusion. The resultant verification rates showed a significant improvement of the three fusion schemes over unimodal and multimodal fusion methods.The developers in [9] applied two optimization techniques to reduce the dimensionality of the iris-signature feature vector. Also, to implement the biometric system, four classifiers and two databases were proposed for more performance evaluation and comparison. The integration of cancellable modalities for Mean-Closure Weighting score-level and Dempster-Shafer based on decision-level fusion for iris and fingerprint were proposed in [10].

2. The proposed method

The six sequential steps that demonstrate the proposed multimodal biometric system are: face-iris image acquisition, feature extraction, classification, score-level fusion, and making the final decision. The four feature extraction methods and their fusion procedure are explained as follows:

2.1 Local binary pattern

The concept of LBP method is based on deference operation between eight neighbours and the center pixel, where the center pixel is considered as a threshold value. If the result of deference operation in equation 1 is zero or positive number, the LBP code is set to one otherwise it is set to zero as as in equations(2)and (3)[11, 12].

$$x_L = g_n - g_c \tag{1}$$

$$s(x_L) = \begin{Bmatrix} 1 & x_L \geq 0 \\ 0 & x_L < 0 \end{Bmatrix} \tag{2}$$

$$LBP_{P,R}(x_c, y_c) = \sum_{p=0}^{p-1} s(x_L) 2^p \tag{3}$$

Where g_n and g_c represent the gray value of neighbours and centre pixel respectively. The two variables x_c and $y_c$are indicated to the coordinates of center pixel. The circle LBP is constructed to provide the flexibility for using different number of neighbours (P) with radius R [12]

2.2 Gray level co-occurrence matrix

The idea of GLCM method is based on the counting of specific pair's pixel value with specific distance and direction angle over the whole image. The GLCM second order features can be extracted in different direction angles, where each angle determines specific relationship such as: 90°, 0°, 45°, and 135°. More than six features can be produced by GLCM, they are: contrast, homogeneity, energy, entropy, correlation, and autocorrelation [13, 14]

2.3 Principle component analysis

The concept of PCA is relayed on computing the eigenvalues and eigenvectors from the covariance matrix where the dimension of feature vector is reduced with maintaining the important values. The covariance matrix, eigenvalues, and eigenvectors can be computedas in equations(4) to (6)[14]:

$$mean\ (\bar{A}) = \frac{1}{n}\sum_{i=1}^{n} a_i \tag{4}$$

$$\mathrm{cov}(a,b) = \frac{\sum_{i=1}^{n}(a_i - \bar{A}_i)\,(b_i - \bar{B}_i)}{n} \tag{5}$$

$$\det\left(\mathrm{cov}_{(a,b)} - \Lambda I\right) = 0 \tag{6}$$

Where $\bar{A}_i$ and $\bar{B}_i$ represent the mean value of vector, while the two parameters a_i and $b_i$refer to the present value of vector a and b. The variable n indicates to the number of rows.

2.4 Fourier descriptors

It is essential method to describe the shape despite its position, rotation, and scale.There are three major steps for implementing the FDs methodas in equations(7) to (9)[5.

$$r(t) = \sqrt{(x(t) - x\ )^2 + (y(t) - y\ )^2} \tag{7}$$

$$x\ = \frac{1}{N}\sum_{t=0}^{N-1} x(t)\ \ ,\ \ y\ = \frac{1}{N}\sum_{t=0}^{N-1} y(t) \tag{8}$$

$$FD_n\ = \frac{1}{N}\sum_{t=0}^{N-1} r(t)\exp\left(\frac{-j2\pi nt}{N}\right) \tag{9}$$

Where the two variables x and y refer to the centred of the shape, while $x(t)$ and $y(t)$ refer to the border points as a complex number. N represents the number of sampling.

2.5 Score Level Fusion

This technique is applied to improve the overall biometric system performance. Firstly, the classification process is performed for face and iris features separately using Euclidean Distance (ED). The second step is normalizing the face and iris scores that are produced from previous step as in equation (10). Third step is fusing the scores using the sum rule as in equation (11)[15]. Figure 1 shows the steps of building a multimodal system based on the score-level fusion technique.

$$S_{Ni} = \frac{S_i - max_{si}}{max_{si} - min_{si}} \tag{10}$$

$$F_{score} = \sum_{i}^{M} S_{Ni} \ (11)$$

Where S_i, S_{Ni}: are the score value and score normalization ofsample *i*, max_{si}, min_{si} represent the maximum and minimum value of the score vector for sample *i*, *M* refers to the number of biometric systems that have been used.

IOP Conf. Series: Materials Science and Engineering **990** (2020) 012021 doi:10.1088/1757-899X/990/1/012021

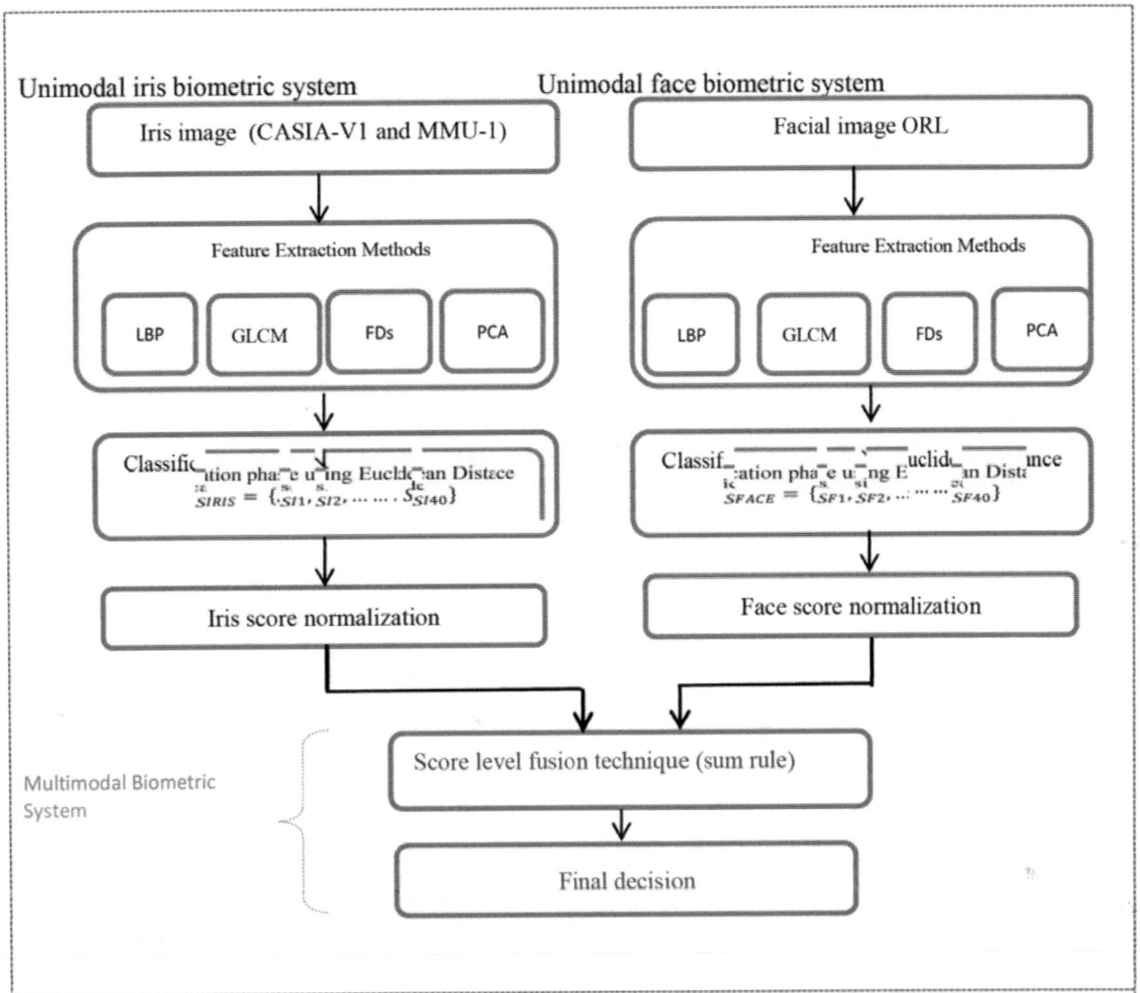

Figure 1. Score-level based multimodal biometric system

3. Results and discussion

The performance results ofthe five different modal systems are tabulated in table 1. The maximum recognition accuracy is 100% and it is achieved by the GLCM, LBP, and PCA methods. The lowest accuracy rate is detected with ORL-MMU-1 dataset which contains blurred iris images. It is clearly noticed from table 1 how much the score-level has improved the accuracy of the unimodal systems, where the rate limits are(95-100)%, these results have advanced the score-level of [8] thathas rate limits (81.17 – 95)% , although [8] implemented its score-fusion system using the right and left iris images with Log-Gabor and LDA feature extraction methods.

Table 1. Performance of unimodal systems and its score-level fusion for 120 subjects

Method	Recognition Rate(%)				
	Three unimodal systems			Two multimodal system	
	ORL	CASIA-V1	MMU-1	ORL-CASIA-V1	ORL-MMU-1
GLCM	100	62.5	60	100	100
LBP	100	95	85	100	95
FDs	90	92.5	50	97.5	95
PCA	92.5	87.5	75	97.5	100

4. Conclusion

The experimental results showed that the score-fusion based multibiometric system has been succeed improving the recognition accuracy of both unimodal systems like face-based recognition and iris-based recognition systems, even with presence of blurred iris images in the ORL-MMU-1 face-iris dataset, the fusion technique satisfied significant accuracy rates (95% and 100%) for classifying 40 subjects. However the GLCM advances all other feature extraction methods by achieving 100% accuracy for both combinational datasets. The score-level fusion of face-iris traits has approved in this work its robustness in classifying the normal and difficult images with promising accuracy rate values.

Acknowledgment

Our thanks to Mustansiriyah Uni. for the support and encouragement

References

[1] Chen C and Chu C 2005 Fusion of Face and Iris Features for Multimodal Biometrics Springer-Verlag BerlinHeidelberg, I-Shou University, Institute of Electrical Engineering
[2] Dhamala P 2012 Multibiometric SystemsM.Sc. Thesis Norwegian University of Science and Technology, Department of Telematics
[3] Srivastava N 2017 Fusion Levels in Multimodal Biometric SystemsEngineering and Technology **6**
[4] Faridah Y, Nasir H and Safie S 2016 Multimodal Biometric Algorithm: A Survey Biotechnology**15**119-124
[5] Hamd M and Mohammed M 2019 Multimodal Biometric System Based Face-Iris Feature Level Fusion(IJMECS) Vol**11**
[6] Sharma D and Kumar A 2011 Multi-Modal Biometric Recognition System: Fusion of Face and IrisFeatures using Local Gabor Patterns International Journal of Advanced Research in Computer Science **2**
[7] Pong K and Lamm K 2014 Multi-resolution feature fusion for face recognition Pattern Recognition**47**556-567, https://dx.doi.org/10.1016/j.patcog.2013.08.023
[8] Sharifi O and Eskandari M 2016 Optimal Face-Iris Multimodal Fusion SchemeSymmetry **8**https://doi.org/:10.3390/sym8060048
[9] Hamd Muthana 2019 Optimized biometric system based iris-signature for human identification IJAIN **5**273-284 https://doi.org/10.26555/ijain.v5i3.407
[10] Dwivedi R and Dey S 2019 A novel hybrid score level and decision level fusion scheme for cancelable multi-biometric verification Applied Intelligence https://doi.org/10.1007/s10489-018-1311-2
[11] Marcel S, Rodriguez Y and Heusch G 2007 On the Recent Use of Local Binary Patterns for Face AuthenticationInternational Journal Of Image And Video Processing Special Issue On Facial Image Processing
[12] Rahim Md, Hossain Md and Azam Md 2013 Face Recognition using Local Binary Patterns (LBP)Global Journal of Computer Science and Technology Graphics & Vision Vol13
[13] Korchiynel R, Farssi S, Sbihi A and Alaoui M 2014 A Combined Method of Fractal and GLCM Featuresfor MRI and CT Scan Images Classification Signal & Image Processing :

IOP Conf. Series: Materials Science and Engineering **990** (2020) 012021 doi:10.1088/1757-899X/990/1/012021

An International Journal(SIPIJ) **5**
[14] ACAR E 2016 Extraction of Texture Features from Local Iris Areas by GLCM and Iris Recognition System Based on KNN European Journal of Technic**6**
[15] Hamd Muthana and Rasool R 2020 Optimized Multimodal Biometric System Based Fusion Technique for Human Identificatio Builtin **9**

IOP Conf. Series: Materials Science and Engineering **990** (2020) 012022 doi:10.1088/1757-899X/990/1/012022

Predicting Surface Hardness of Commercially Pure Titanium Under Plasma Nitrocarburizing Based on Experimental Data

A S Darmawan[1], W A Siswanto[1], P I Purboputro[1], B W Febriantoko[1], T Sujitno[2], A Hamid[3]
[1]Department of Mechanical Engineering, Faculty of Engineering, Universitas Muhammadiyah Surakarta, Pabelan, Surakarta 57162, Indonesia
[2]National Nuclear Energy Agency (BATAN), Yogyakarta, Indonesia
[3]Department of Mechanical Engineering, Faculty of Engineering, Universitas Batam, Riau, Indonesia

E-mail: Agung.Darmawan@ums.ac.id

Abstract. One of the techniques to improve the surface hardness of commercially pure titanium is by using plasma nitrocarburizing. This paper presents an equation to predict the surface hardness with the variable of the treatment temperature and process time of plasma nitrocarburizing. Experimental studies were first conducted to collect surface hardness at different elevated temperatures with different process time. The prediction equation was then developed by using Lagrange's Multivariable Interpolation method. It is found that commercially pure titanium surface hardness can be predicted as a function of time and temperature. This equation is able to predict the surface hardness of the experimental data with high accuracy. The errors of the predicted hardness to the experimental results are less than 1 percent.

1. Introduction

Titanium has outstanding properties in term of high strength to weight ratio, highcorrosion resistance and excellent biocompatibility [1-5]. Nowadays, titanium is widely used in medical and industry field [6-9].Titanium is widely used in medical field because of its excellent biocompatibility and high corrosion resistance. Its applications involve surgical implement and implants, such as hip ball and socket in hip joint replacement. Titanium application is also in other several medical fields such as bone fitting, dental implant materials, the retaining structure of the heart valves and replacement of the skull [10-12]

Although the hardness properties are considered as outstanding, however when the applications involve sliding conditions, the surface has poor tribological properties. Therefore, when titanium is used as a part of which will experience friction with other parts, for example in hip joint replacement application, the improvement of the hardness value of the surface will be required [13].

The problem occurs when the hardness is increased by a certain treatment to the whole part, there is a possibility the material will become brittle, which is not expected. Any excessive loading applied to the material will damage it easily. Increasing the hardness at the surface without changing the ductility

[1] Agung.Darmawan@ums.ac.id.

Content from this work may be used under the terms of the Creative Commons Attribution 3.0 licence. Any further distribution of this work must maintain attribution to the author(s) and the title of the work, journal citation and DOI.
Published under licence by IOP Publishing Ltd

properties of materials on the inside of the material will increase the toughness properties of material without changing the ductility in the inner part. This process is referred to as surface hardening.

One of the surface hardening processes is nitrocarburizing. Nitrocarburizing is basically a thermo chemical process in which a process of diffusion of nitrogen and carbon atoms toward the surface of metallic materials at certain elevated temperatures. Heat is required to enhance the diffusion of hardening species into the material's surface. Nitrocarburizing can be applied to liquid, solid, and plasma atmospheres. This is widely used to increase the hardness value in the surface of steel [14, 15].

The plasma nitrocarburizing produces faster nitrogen and carbon diffusion, more friendly with environment, more economical, and, lower gas consumption compared to other nitrocarburizing techniques [16]. The surface hardness improvement as a result of plasma nitrocarburizing process is significantly influenced by the heating temperature and treatment process time. In order to reduce the time and cost in investigating the proper treatment setting of heat and process time to get the expected surface hardness, it is necessary to have an empirical prediction formula for determining the surface hardness as a function of the heat temperature and the treatment process time.

This current research searches the mathematical formula of the surface hardness as a function of heat and time for nitrocarburizing based on experimental results.In this research, the commercially pure (cp) titanium is used. The surface hardness improvement by plasma nitrocarburizing is conducted in various settings of heat and time then the surface hardness is measured. The prediction formula is developed by using Lagrange multivariate interpolation. The formula is derived by considering all experimental data obtained from the plasma nitrocarburizing experiments.

2. Materials and Methodology

The chemical composition of cp titanium is as follows: N: 0.04%, C: 0.05%, H: 0.003%, Fe: 0.13%, O: 0.11%, Al: 0.49% S: 0.03, Ti: balance. Before the nitrocarburizing surface treatment, the hardness value of the material is 53.99 HV.

In the preparation in plasma nitrocarburizing, cp titanium material is cut with the size of 1 cm × 1 cm × 0.3 cm for each specimen. There are six specimens prepared for this purpose. Three specimens are used in the process of plasma nitrocarburizing at 350°C and three specimens are used at 450°C. The surfaces of the specimens are then grinded and polished using polycrystalline diamond until they are all clean and shiny.

In plasma nitrocarburizing process, the plasma is formed in a vacuum by means of high voltage electrical energy in which the positive ions of nitrogen, carbon and hydrogen are accelerated to strike the cathode. The work piece is maintained at a negative DC (direct current) high voltage source of 250-850 volts in the presence of an electric field.

The gases are separated, ionized, and accelerated toward the work piece (cathode). The kinetic energy of ion is converted into heat energy by ion bombardment. This energy not only heats the work piece but also implants ions directly and produces the cathode sputtering. The electrons are forced out from the surface of the work piece. Some of the ions implanted into the surface of the specimen, the other led to the cathode sputtering. Furthermore, absorption and diffusion of nitrogen and carbon atoms led to the formation of the compound layer.

After the plasma nitrocarburizing processes are completed, each specimen is tested of its hardness using micro Vickers hardness tester. This hardness test follows the standard ASTME384.

Many researchers have implemented Lagrange interpolation in their works and also published [17-19]. Basically Lagrange give a set of $k + 1$ data points, $(x_0,y_0),\ldots,(x_j,y_j),\ldots,(x_k,y_k)$, where no two variables x_j are the same, the interpolation polynomial in the Lagrange formula is a linear combination.

$$L(x) = \sum_{j=0}^{k} y_j \, l_j(x) \tag{1}$$

of Lagrange basis polynomials

$$l_j(x) = \prod_{\substack{0\le m\le k \\ m\ne j}} \frac{x - x_m}{x_j - x_m} = \frac{(x - x_0)}{(x_j - x_0)} \cdots \frac{(x - x_{j-1})(x - x_{j+1})}{(x_j - x_{j-1})(x_j - x_{j+1})} \cdots \frac{(x - x_k)}{(x_j - x_k)} \quad (1)$$

where $0 \le j \le k$. Note how, given the initial assumption that no two x_i are the same, $x_j - x_m \ne 0$, so this expression is always well-defined. The reason pairs $x_i = x_j$ with $y_i \ne y_j$ are not allowed is that no interpolation function L such that $y_i = L(x_i)$ would exist; a function can only get one value for each argument x_i. On the other hand, if also $y_i = y_j$, then those two points would actually be one single point.

For all $j \ne i$, $l_j(x)$ includes the term $(x - x_i)$ in the numerator, so the whole product will be zero at $x = x_i$:

$$l_{j\ne i}(x_i) = \prod_{m\ne j} \frac{x_i - x_m}{x_j - x_m} = \frac{(x_i - x_0)}{(x_j - x_0)} \cdots \frac{(x_i - x_i)}{(x_j - x_i)} \cdots \frac{(x_i - x_k)}{(x_j - x_k)} = 0 \quad (3)$$

on the other hand,

$$l_i(x_i) := \prod_{m\ne i} \frac{x_i - x_m}{x_i - x_m} = 1 \quad (4)$$

All basis polynomials are zero at$x = x_i$, except $l_i(x)$, for which it holds that$l_i(x_i)$, because it lacks the $(x - x_i)$ term.It follows that$y_i l_i(x_i) = y_i$, so at each pointx_i,

$$L(x_i) := y_i + 0 + 0 + \cdots + 0 = y_i \quad (5)$$

showing that L interpolates the function exactly.

Lagrange interpolation can be used to develop a multi-variable formula [20]. Let $f = f(x_1, \ldots, x_m)$ is an m-variable multinomial function of degree n. Since there are $\rho = \binom{n+m}{n}$ terms in f, it is a necessary condition to have ρ distinct points $(x_{1,i}, \ldots, x_{m,i}, f_i) \in R^{m+1}$, $1 \le i \le \rho$, $f_i = f(x_{1,i}, \ldots, x_{m,i})$, for f to be uniquely defined. This means,

$$f(x_1, \ldots, x_m) = \sum_{e_i \cdot 1 \le n} \alpha_{e_i} \mathbf{x}^{e_i} \quad (6)$$

where the $\alpha_{e_i}$is the coefficients in f, $\mathbf{x} = (x_1, \ldots, x_m)$ is the m independent variables of f, $e_i = (e_{1i}, \ldots, e_{mi})$ is an exponent vector with nonnegative integer entries consisting of an ordered partition of an integer between 0 and n inclusive, $e_i.1 = \sum_{j=1}^{m} e_{ji}$ is the usual vector dot product, and $\mathbf{x}^{e_i} = \prod_{j=1}^{m} x_j^{e_{ji}}$.

3. Results and Discussion

The surface hardness of cp titanium resulted from plasma nitrocarburizing process are illustrated in Fig. 1. For specimens of the plasma nitrocarburizing processes at temperatures of 350 ° C, there is a significant increasing in hardness value when the process time is extended from 180 minutes to 240 minutes. The hardness can be improved as high as 24.4%. The hardness improvement, however does not show a linear behaviour when the process time is extended. When the process time is extended to 300 minutes, the hardness value is increased only 2.3%. The hardness improvement is not significant when it is compared with that of process time extension from 180 minutes to 240 minutes.

The similar trend appears for the plasma nitrocarburizing processes at temperatures of 450 ° C as well. There is a significant increasing in hardness value when the process time is extended from 120 minutes to 180 minutes as high as 17%. However, after the process time is extended from 180 minutes to 240 minutes, the increasing hardness is only 4%.

Figure 1 also shows that hardness value of specimens that are resulted from the plasma nitrocarburizing process at temperature of 450 °C is higher compared with specimens that are processed at temperature of 350 °C.

The results of hardness test showed that the higher the temperature, the hardness value will be higher, and the longer the process time, the higher the hardness value. This is because the diffusion of carbon and nitrogen atoms depends on the process time and temperature. The higher the time and temperature, the more diffusion of carbon and nitrogen atoms and the distance is longer as well. Furthermore, the layer compound Ti (N, C) which is formed also thicker so the hardness become higher.

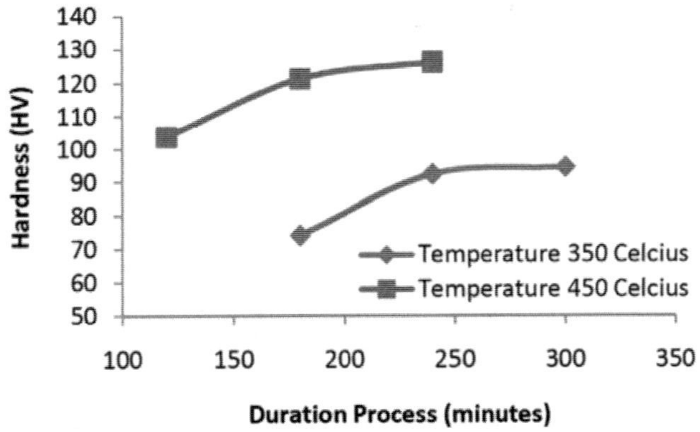

Figure 1. The surface hardness of cp titanium that is processed plasma nitrocarburizing at temperatures 350 °C and 450 °C.

The Lagrange polynomial interpolation is solved by using inverse method implemented in FreeMat. The complete equation of the prediction:

$$\begin{aligned} Hp(t,\, T) = {} & 0.000000001591435t^2T^2 - 0.000008314814 81t^2T - 0.00125182291667t^2 \\ & - 0.000006486805 56tT^2 + 0.001129444444 45tT + 1.63007812499982t \\ & + 0.00066025000000T^2 + 0.530999999999990T \\ & - 342.42062499998030 \end{aligned} \tag{1}$$

The process time range is from 120 minutes to 300 minutes while the temperature range is from 350 °C to 450 °C.From the equation, the highest surface hardness is 127.11 HV at process time 228 minutes and temperature 450 °C. The smallest surface hardness is 40.29 HV at process time 120 minutes and temperature 350 °C. Figure 2 shows the surface hardness curve of cp titanium as a function of process time and temperature that is resulted from this multivariate interpolation method.It can be shown from Table 1 that the predicted surface hardness has high accuracy (error is less than 1%) if the decimal digits of significant are 9 or more.

IOP Conf. Series: Materials Science and Engineering **990** (2020) 012022 doi:10.1088/1757-899X/990/1/012022

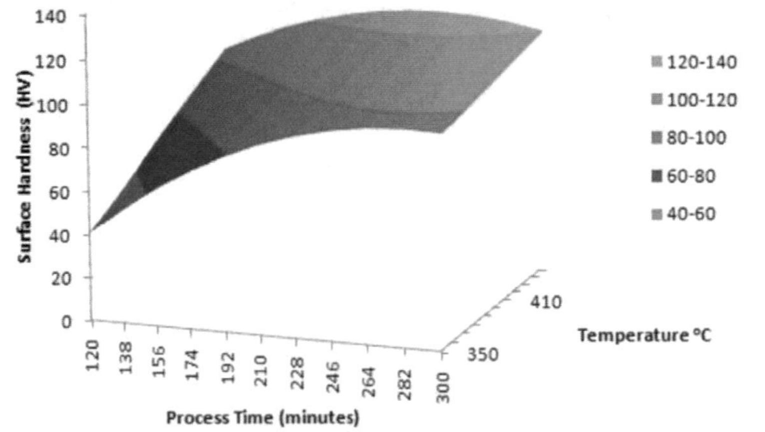

Figure 2. The surface hardness of cp titanium that is processed plasma nitrocarburizing at temperatures 350 °C and 450 °C.

Table 1. Error in surface hardness value due to the number of decimal digits

Process Time (Minutes)	Temperature (°C)	Experiment Results (HV)	Error (%)			
			14 Decimal Digits	10 Decimal Digits	9 Decimal Digits	8 Decimal Digits
180	350	74.16	9.97E-06	0.08	0.45	21.69
240	350	92.25	1.42E-07	0.11	0.64	31.03
300	350	94.41	2.16E-07	0.17	0.99	47.41
120	450	103.7	5.28E-08	0.04	0.24	11.38
180	450	121.31	1.00E-07	0.08	0.46	21.94
240	450	126.17	1.71E-07	0.13	0.78	37.54

4. Conclusion
Nitrocarburizing plasma increased the surface hardness of commercially pure titanium. By using the multivariate Lagrange interpolation method, a prediction equation is obtained:

$$\begin{aligned} Hp(t,\ T) = {} & 0.00000001591435t^2T^2 - 0.00000831481481t^2T - 0.00125182291667t^2 \\ & - 0.00000648680556tT^2 + 0.00112944444445tT + 1.63007812499982t \\ & + 0.00066025000000T^2 + 0.53099999999990T \\ & - 342.42062499998030 \end{aligned}$$

This equation is accurate with deviations less than 1%.

Acknowledgment
The authors gratefully acknowledge financial support from Universiti Tun Hussein Onn Malaysia under Fund Scheme GIS Vot 0809. The authors also would like to thank BATANYogyakarta Indonesia for plasma nitro carburizing processes and Mechanical Engineering Laboratory of Universitas Muhammadiyah Surakarta for mechanical testing.

References
[1] YuH Y, CaiZ B, ZhouZ R and ZhuM H 2005*Wear***259** 910.

[2] TanM J and ZhuX J 2007*Archives of Materials Science and Engineering***28** 5.
[3] Darmawan A S, Siswanto W A, Sujitno T 2013*Applied Mechanics and Materials***315** 700.
[4] DongH and BellT 2000 *Wear***238** 131
[5] Darmawan A S, Siswanto W A, Sujitno T 2013 *Research Journal of Applied Sciences, Engineering and Technology* 5(4) 1351.
[6] FuY, and BatchelorA W 1998*Wear***214** 83.
[7] HamadaK, KonM, HanawaT, YokoyamaK, Miyamotoand AsaokaK 2002*Biomaterials***23** 2265.
[8] LuoY and GeG 2009 *Tribol. Int.***42** 1373.
[9] TrticaM, GakovicB, BataniD, DesaiT, PanjanP and RadakB 2006 *Appl. Surf. Sci.***253** 2551.
[10] LiuX, ChuP K and DingC 2004*Mater. Sci. Eng. R***47** 49.
[11] LiuX, PoonR W Y, KwokS C H, ChuP K and DingC 2004*Surf. Coat. Technol.***186** 227.
[12] TianY S, ChenC Z, LiS T and HuoQ H*Appl. Surf. Sci.***242** 177.
[13] Darmawan A S, Siswanto W A, Febriantoko B W, Hamid A and Sujitno T2019 *Materials Science Forum***961** 97.
[14] BellT, SunY and SuhadiA 2000*Vacuum***59**14.
[15] KrishnarajN, SrinivasanP B, IyerK J L and SundaresanS 1998*Wear***215** 123.
[16] KarakanM, AlsaranA and CelikA 2004*Mater. Des.***25** 349.
[17] AmatS, BusquierS, EscuderoA and TrilloJ C 2008*J. Comput. Appl. Math.***221** 47.
[18] ChengZ and MarkJ W 2010*Signal Process.***90** 2749.
[19] OccorsioD and RussoM G 2014*J. Comput. Appl. Math.***259** 24.
[20] LiangX Z, WangR H, CuiL H, ZhangJ L. and ZhangM 2006*J. Comput. Appl. Math.***195** 192.

Analysis of Characteristics of Activated Carbon from Cacao (*Theobroma cacao*) Skin Waste for Supercapasitor Electrodes

YYetri[1*], Mursida[2], DDahlan[2], Muldarisnur[2],E Taer[3],Febrielviyenti[4]
[1]Mechanical Engineering Department, Politeknik Negeri Padang, Padang, Indonesia
[2]Physical Department, Andalas University, Padang, Indonesia,
[3]Physical Department, Riau University, Pekanbaru, Indonesia,
[4]Central Laboratory of Riau Islamic University.

E-mail: yuliyetri@pnp.ac.id

Abstract. The production of activated carbon from the cacao (*Theobroma cacao*) skin waste was investigated to be used as a super capacitor electrode. This study was conducted in order to obtain the characteristics of the activated carbon electrodes of cacao skins. Activated carbon electrodes were made by a combination of chemical and physical activation methods. Samples were pre-carbonized at 250°C for 2.5 hours, whereas chemical activation was performed using KOH activators of 0.3M and 0.4M, with physical activation temperatures of 700°C, and the particle size of the activated carbon used is smaller than 38 μm. Then it was characterized by X-ray diffraction (XRD), the CiklicVoltammetric (CV) method, and Scanning Electron Microscopy (SEM). Hence, the analysis was carried out on the specific dimensions, density and capacitance. It was found that the mass, diameter, thickness, and density of the electrodes have decreased, and the specific capacitance of the electrodes has increased along with the increase in KOH concentration from 0.3 M to 0.4 M. The specific capacitance value at a concentration of 0.3 M was 90.2 F/gr with a density of 0.850gr/cm^3 and a concentration of 0.4 M was 140.2 F/gr with a density of 0.802 gr/cm^3. The X-ray diffraction curve data showed that the value of Lc and the lattice distance (d_{002}) of C0.4 samples was smaller than the C0.3 sample. From the obtained results, the surface area of the C0.4 sample was greater than the C0.3 sample. It was clear that the chemical activation of 0.4M produces electrodes with better performance than 0.3M.

1. Introduction

Indonesia is the third country in the world in cacao producing besides Ghana and Ivory Coast. Cacao is one of the productive plants that produces fruit and can bear fruit throughout the year [1]. Cacao fruit consists of three main components, namely fruit skin, placenta, and seeds (Figure 1). Fresh cacao skins consist of 73% skins, 24.2% seeds, and 2% placenta [1-3], and makes the cacao skins becoming the most waste of all cacao plants, which is about 75% of its fresh skins [4]. One ton of dried cacao beans is equivalent to 10 tons of fresh cacao skins. To overcome the high production number, effective handling of waste is needed so that it does not interfere with the production process and the environment around the processing plant.

[1]yuliyetri@pnp.ac.id

Content from this work may be used under the terms of the Creative Commons Attribution 3.0 licence. Any further distribution of this work must maintain attribution to the author(s) and the title of the work, journal citation and DOI.
Published under licence by IOP Publishing Ltd

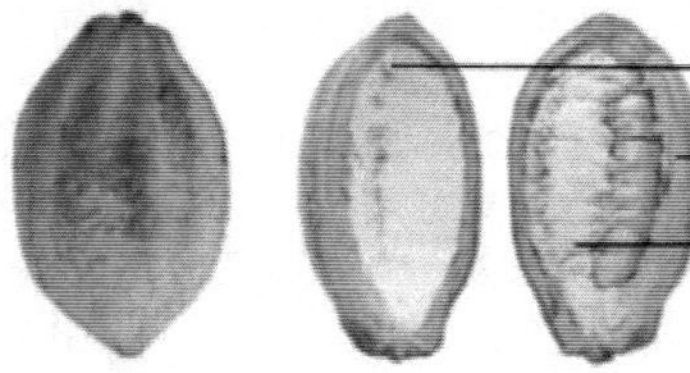

Figure 1. Cacao fruit [5]

Various alternatives have been made to deal with the final disposal of cacao fruit waste. One of them is creating valuable products, such as antioxidants in cosmetic products [6], additives [3], as animal feed [7], adsorbents of Rhodamin B dyes [8], absorbent heavy metals [9-10], corrosion inhibitors [11], restoration of mechanical properties of mild steel after corrosion [5], activated carbon [12], absorbent amoxicillin [13], paper production [14], and many others.

Biomass materials can be used as raw material to make activated carbon, because biomass is composed of lignocellulose, cellulose, and hemicellulose which are organic carbon. Activated carbon can be made from all carbon-containing materials, both organic and inorganic carbon with the condition that the material has a porous structure and meets the requirements of the Indonesian National Standard (SNI) No. 06-3730-1995. Some of utilized biomass such as wood, lignite, bones, palm shells, oil palm bunches, agricultural waste such as coffee fruit skins, cacao fruit husks, rice husks, straw, cob and cornstalks, fruit skins, fruit seeds, grain skins, and others [14].

Cacao skin is one of the biomass wastes that meets the requirements to become activated carbon. This research was set up to make activated carbon from cacao shell using KOH activator with variations in concentration at a temperature of 700°C. By processing cacao skin waste into carbon electrodes as a basic material for super capacitors, not only is the environment protected from pollution, but it will also reduce the number of carbon electrodes imported by industry in Indonesia.

2. Research Method

The making of activated carbon from biomass cacao skin until printed into pellets has been described in previous publications [15]. Carbon pellet moulding was carried out using a hydraulic press at 8 tons pressure. The next step was the carbonization process using furnaces, starting from room temperature to 305°C, and held for one hour. Then proceed to a maximum temperature of 600°C in the N_2 gas environment and continued to physics activation at 700°C using CO_2 gas. The 305°C temperature resistance was taken from durian skin' temperature resistance, due to the similar texture of cacao fruit and durian skin after experiencing the dehydration process [16]. Dimension measurements carried out include the measurements of mass, diameter, thickness of each pellet prior to and after the physics' carbonization-activation [15].

After measuring the density, the carbon pellet sample was polished to a diameter of ± 8 mm and a thickness of ± 0.3 mm, then washed until the pH of the washing water became neutral. Then neutral carbon electrodes were immersed in 1 M sulphuric acid (H_2SO_4) solution which functioned as an electrolyte in super capacitor cells.

The electrochemical properties of carbon electrodes were tested by assembling a super capacitor cell. The specific capacitance value of the super capacitor can be calculated using the CiclicVoltamentry (CV) method at an induced rate of 1 mV/s to 100 mV/s [15].

3. Results and Discussions

3.1 Physical properties

During the drying process, the skin of cacao fruit experienced a reduction in water content by 84, 47%, where the water content of fresh cacao skin was 70.582%, and became 10.964% when it dried The carbonization process causes the decomposition of organic material in cacao shells, and releases volatile substances, where most of the non-carbon elements will be released into the air [17]. The space left by these non-carbon elements forms a pore, nevertheless the pore volume and surface area formed is usually small below the standard of activated carbon. As a result of the release of volatile elements, the carbon produced has decreased. The percentage of mass depletion of cacao shells to carbon ranges between (19.3-34.2)%. The physical observations of carbon from cacao shells to carbon pellets can be seen in Figure 2.

The low water content because of the smaller surface of activated carbon contains fewer polar functional groups so that the interaction between polar water vapour is also small [18-19]. Low level of this water content also indicates that the substance evaporates, and other compounds in the activated carbon the skin of cacao is more easily released, so that the surface area of activated carbon is getting bigger and the pores of charcoal are increasing.

The measurement results of mass dimensions (m), diameter (d), thickness (t), and density (ρ) in each variation, before and after carbonization-activation of physics can be seen in Table 1, and 2. The increase on KOH concentration from 0 3 M to 0.4 M stimulated a decrease in the dimensions and density of carbon pellets. These results were in accordance with previous studies conducted by ErmanTaer (2016) on the manufacture of super capacitor electrodes from rubber wood using chemical activation of KOH and HNO_3. The reduction in density on the carbon electrode is due to the addition of the pore structure from the combination of activation [20]. Analysis of the dimensional shrinkage and density of activated carbon in both variations is presented in Tables 1 and 2.

Figure 2. Preparation of activated carbon electrodes for cacao skin husks(a) cacao skin skin, (b) after dehydration, (c) after pre-carbonization, (d) after sifting, (e) after printing, (f) pellets after carbonization and activation, and (g) pellets after polishing.

Table 1. Percentage of the electrodes shrinkage with a 0.3M KOH activator at temperature of 700°C

No	Variable	Beginning	Ending	Losing	Shrinking (%)
1	Massa (gram)	0.703	0.256	0.447	63.6
2	Diameter (cm)	1.985	1.480	0.505	25.3
3	Thick (cm)	0.254	0.187	0.067	26.9
4	Density (gram/cm^3)	0.918	0.850	0.038	9.9

Table 2. Percentage of the electrodes shrinkage with a 0.4M KOH activator at temperature of 700°C

No	Variable	Beginning	Ending	Losing	Shrinking (%)
1	Massa (gram)	0.689	0.231	0.458	64.1
2	Diameter (cm)	1.978	1.486	0.492	29.8
3	Thick (cm)	0.240	0.187	0.763	25.0
4	Density (gram/cm^3)	0.950	0.802	0.148	18.6

The mass loss occurring in the 0.4 M sample with activation temperature of 700°C was greater than the mass loss in the 0.3 M sample with the same activation temperature, however it was not too significant. It is because the carbonization process causes the decomposition of organic material in the cacao skin and releases volatile substances such as water vapour during the carbonization and activation process. The dominant mass loss will impact on the diameter and thickness. There was a significant difference in the thickness dimension, the shrinkage of thickness in the 0.3 M sample with a temperature of 700°C which was much greater than in the 0.4 M sample with the same temperature. On the other hand, the diameter dimension showed an adjacent value. The physical parameter values would clearly produce lower densities in carbon after activation than carbon before carbonization [13].

The mass shrinkage was high in number caused by the evaporation of material other than carbon during the carbonization and activation process. Whereas the shrinkage in diameter and thickness was caused by heating which directly presented the empty space between carbon atoms, so that bonds or carbon atoms form stronger bonds [21]. Shrinkage of mass was more dominant than shrinkage of diameter and thickness, this condition would clearly produce a lower density on activated carbon than green carbon [21].

3.2 XRD analysis

X-ray diffraction analysis (XRD) utilized a 2θ diffraction angle that was in the 10°-100° angle range. Previous publications have explained the pattern of XRD activated carbon of cacao skin with concentrations of 0.3M and 0.4M KOH activators [15].

The difference in the electrode structure of 0.3 M and 0.4 M at 700°C could be observed from the peak height and width half the height of the the peak of the XRD diffractogram which were 95.380 nm and 15.176 nm for the 0.3 M sample with a temperature of 700°C. As for the 0.4 M sample with a temperature of 700°C were 194.64 nm and 6.8725 nm [15]. The shift in peak height and width half the height of the peak indicated the rearrangement of carbon atoms.

The XRD data fitting results from each sample were obtained at the lattice distance d_{002} and d100, as shown in Table 4. The largest Lc / La ratio number was obtained on the 0.3M electrode sample and the smallest on the 0.4M electrode sample at the same temperature 700°C .

Table 3. Results of data fitting of X-ray diffraction data of cacao skin electrodes

Kode Elektroda	$2\theta_{(002)}$ (°)	$2\theta_{(100)}$ (°)	$d_{(002)}$ (Å)	$d_{(100)}$ (Å)	Lc (Å)	La (Å)
C 0,3 M	23.569	44.781	3.771705	2.022225	20.01669	11.93877
C 0,4 M	24.747	44.634	3.594769	2.028543	3.256256	30.3769

Based on the empirical formula used, to produce a higher surface area of carbon electrodes a smaller lattice distance (d_{002}) and smaller dimensions of microcrystalline (Lc) are required [15]. The X-ray diffraction curve data in Table 4 shows that the C 0.4M sample has a smaller Lc value and the lattice distance (d_{002}) than the C 0.3M sample, so it can be indicated that the surface area of the C 0.4M sample is greater than sample C 0.3M.

3.3 Scanning Electron Microscopy (SEM) analysis

The scanning electron microscope (SEM) analysis aims to determine the morphology of the carbon electrodes of cacao skins by looking at the distribution of the portions. The sample observations in Figure 3 were carried out at 5000 X magnification.

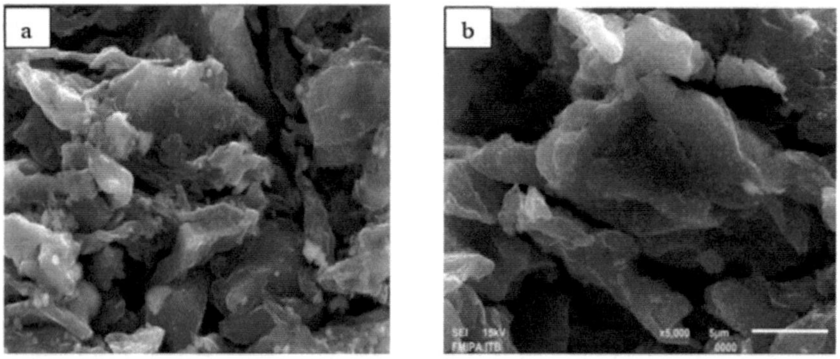

Figure 3. Observation of SEM carbon electrode temperature of 700°C, magnification of 5000 Xa) 0.3M, and b) 0.4M.

Figure 3a shows the surface morphology of the activated carbon of the cacao skins. Pores between particles are spread almost evenly on the surface of the sample, some have elongated and irregular shapes. There are still small particles on the surface of the sample which indicate the presence of material other than carbon in the 0.3M 700°C sample or free particles that do not evaporate completely during the carbonization and activation process. The particles are clearly more visible and there are pores in between these particles that are spread evenly with various shapes and sizes.

In Figure 3b shows the morphology of 0.4M carbon electrode 700°C, where the pores between particles are larger than the carbon electrode sample 0.3M 700°C, with irregular size and shape. It is also seen chunks on the surface of the carbon electrode which shows a denser electrode sample structure. There are particles formed more clearly having quite large pore shaped with uneven size in between them. Compared with the morphology of durian skin carbon electrodes with the same magnification and physical activation time, it shows that the carbon particles are relatively dense, elongated and irregular in shape, and the pores between the particles have formed. Closed pores cause the density of carbon electrodes to increase, resulting in small specific capacitance [20].

3.4 X-ray Dispersive Energy (EDX) analysis

Composition analysis with X-ray energy dispersion (EDX) represents the percentage of elemental content contained in carbon electrode samples, as seen in Table 4.

Table 4. Content of compounds in carbon electrode samples

Contents	C 0,3M 700°C		C 0,4M 700°C	
	Massa %	Atom %	Massa %	Atom %
C	87.05	91.02	87.87	91.49
O	10.12	7.95	9.65	7.54
Mg	0.75	0.39	0.94	0.48
Ca	2.07	0,65	1.55	0.48
Total	100%			

Table 4 shows that the elements contained in carbon electrode samples consist of Carbon, Oxygen, Magnesium and Calcium. At the activation of 0.3 M and 0.4 M, the percentage of carbon element content is higher than other elements, namely 91.02% at 0.3 M and 91.49% at 0.4 M. The high content of C indicates that carbon electrode samples has a very high purity. Element O has the second highest percentage after element C, and occurred when carbonization of oxygen content in carbon samples is not completely decomposed or bonds are formed in the activation process [11]. Ca content is presence because it is found in the composition of the skin of cacao fruit. Mg element is indicated to derive from unclean steel balls when ball milling.

The quantity of element C in the 0.4 M carbon electrode sample at 700°C temperature is greater than that of the 0.3 M carbon electrode sample at the same temperature. This specifies that the 0.4 M KOH sample contains more carbon than the 0.3 M KOH sample. The addition of KOH activator concentration from 0.3 M to 0.4 M results in the decomposition of non-carbon content, so that the element C produced is higher.

3.5 Electrochemical properties

Cyclic Volta metric (CV) measurement data is stated in the specific capacitance curve (Csp) of the voltage in the scan rate of 1 mV/s. The CV curve demonstrates the graph of current strength against voltage at KOH concentrations of 0.3 M and 0.4 M, as shown in Figure 4.

Figure 4 points out the CV curve for 0.4 M KOH treatment having a better capacitive response than the 0.3 M KOH sample. This is characterized by a wider middle potential area, so as to increase the number of pores of carbon electrodes. The increase in the number of pores is in line with the increase in the number of ions flowing into the electrodes marked by the width of the charge-discharge curve, as shown in Figure 4. The wider the charge-discharge curve, the greater the specific capacitance produced by the activated carbon electrode.

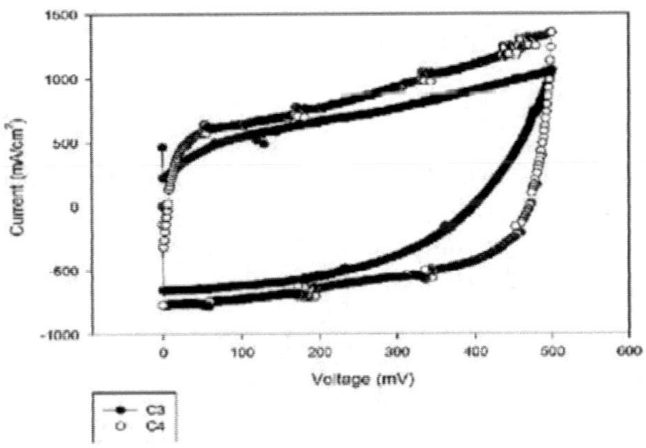

Figure 4. CV curve of activated carbon with KOH 0.3 M and 0.4 M.

0.4 M carbon electrode sample temperature of 700°C has a better capacitive response, this is characterized by a wider central potential area, a greater charge process at 0 V and a horizontal discharge curve shape when compared with the sample 0.3 M 700°C. The shape of the CV curve of 0.4 M 700°C is square, known as the most ideal form for activated carbon electrodes. An increase in the value of specific capacitance (Csp) improves as the increase of KOH concentration from 0.3 M to 0.4 M, as seen in Figure 5. It is illustrated within the value of specific capacitance (Csp) at concentrations of 0.3 M and 0.4 M with an average mass of 0.0132 and 0.0112 is 90.2 F/gr and 140.2 F/gr.

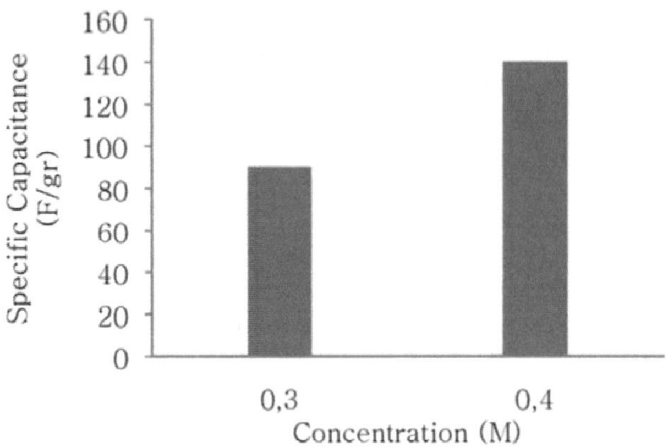

Figure 5. Comparison diagram of specific capacitance values based on variations in KOH activator concentration, at an activation temperature of 700°C..

4. Conclusions

The activated carbon of cacao skins has the potential to be used as a basis for super capacitor electrodes. Analysis of dimensions, density and specific capacitance on the cacao skin carbon electrodes with variations in concentration showed the diameter and thickness of the carbon electrode pellets decreased. Shrinkage of mass and an increase in KOH concentration result in a decrease in the density value of carbon electrodesnamely 0.850 gr/cm^3 for C 0.3M and 0.802 gr/cm^3 for C 0.4M. XRD analysis indicates that the surface area of the C sample of 0.4 M is greater than the C sample of 0.3M, because the 0, 4M C sample has smaller Lc value and lattice distance (d_{002}) than the C 0.3 sample. The specific capacitance value (Csp) increases in line with the increase on KOH concentration from 0.3M to 0.4M, which is from 90.2 F/gr to 140.2 F/gr.

Acknowledgement

The author would like to thank Ministry of Education, Republic of Indonesia in providing the research funding through the HibahPenelitian Dasar (Basic Research Grant) with contract Number: no: 401/PL9.15/PG/2020. The appreciation also goes to Department of Physics, University of Riau for the assistance in characterizing activated carbon of cacao skin.

References

[1] Yetri Y and Sukatik 2017*Orient. J. Chem.***33(4)** 2071-79
[2] Taer E, Mustika W S and Taslim R 2016*ProsidingSeminar Nasional Fisika SNF2016***V**49-54
[3] Tissos N P, Dahlan D and Yetri Y2018*Int. J. Adv. Sci. Eng. Inf. Technol.***8(4)**1290-95
[4] Cruz G, Pirila M, Huuhtanen M, Carrion L, Alverengga E and Keisi R L2012*Conf. Proceedings 1554* (Peru: Universidad Nacional de Tumbes)
[5] Yetri Y, Emriadi, Jamarun N and Gunawarman 2016*Rasayan J. Chem.***9(4)**716-727
[6] Karim A A, Azlan A, Ismail A, Hashim P and Abdullah A N2014*Malays. Cocoa* .**8** 49-56
[7] Marcel B K G, André K B, Théodore D and Sérapin K-C 2011*Int. J. Agron. Agric. Res.* **1(1)** 9-19
[8] Purnawati H and Utami B 2014*Skripsi*(Surakarta: UniversitasSebelasMaret)
[9] Masitoh Y F and Sianita M M 2013*UNESA J. Chem.***2(2)**23-25
[10] Baroroh A, Moelyaningrum A D and Ellyke2017(Jember: UNEJ PRESS)
[11] Wiaya M and Wiharto M 2017*Jurnal Kimia dan Pendidikan Kimia***2(1)** 66-71
[12] Brebu M and Vasile C 2010*Cell. Chem. Technol.***44(9)**353-363

IOP Conf. Series: Materials Science and Engineering **990** (2020) 012023 doi:10.1088/1757-899X/990/1/012023

[13] Tejada C N, Almanza D, Villabona A, Colpas F and Granados C2017*Ingeniería y Competitividad***19(2)**45-54
[14] Daud Z, Kassim A S M, Aripin A M, Awang H and Hatta M Z M 2013*Aust. Basic &Appl Sci.* **7(9)**406-411
[15] Yetri Y, Mursida, Dahlan D, Taer E, Agustino and Muldarisnur2020*Key Eng. Mater.***846** 274-281
[16] Taer E, Dewi P, Sugianto, Syech R, Taslim R, Salomo, Susanti Y, Purnama A, Apriwandi, Agustino and Setiadi R N 2018*AIP Conference Proceedings 1927*
[17] Budianto A, Romiarto and Fitrianingtyas 2016*Prosiding Seminar Nasional Sains dan TeknologiTerapan IV* 207-212
[18] Kierzek K, Frackowiak E, Lota E, Grylewicz G and Machnikowski J2004*Electrochim. Acta* **49**515-523
[19] Fauziah N2009*Skripsi*(Bogor: InstitutPertanian Bogor)
[20] Afrianda A2018*Skripsi*(Pekanbaru: Universitas Riau)
[21] Muchammadsam I D, Taer E and Farma R 2015 *JOM FMIPA* **2(1)**

IOP Conf. Series: Materials Science and Engineering **990** (2020) 012024 doi:10.1088/1757-899X/990/1/012024

Material Density of Composite hydroxyapatite Bovine Bone-Borosilicate formed by Compaction and Sintering Techniques

Burmawi[1], A Syahroom[2], N Jamarun[3], S Arief[3], Gunawarman[4]
[1] Mechanical Engineering, Universitas Bung Hatta
[2] Mechanical Engineering, University Teknologi Malaysia
[3] Chemistry, Andalas University
[4] Mechanical Engineering, Andalas University

E-mail: burmawi@bunghatta.ac.id

Abstract. Hydroxyapatite Ca10 (PO4) 6 (OH) 2 is a calcium compound whose chemical composition is similar to human bone minerals. In this study, hydroxyapatite was formed into a composite material by adding borosilicate (borax and silica). For the formation of this composite material, the Hydroxyapatite borosilicate composition has been adjusted, namely: 90:10, 85:15, 80:20, 75:25, 70:30 wt.%. The samples were made with different compacting forces, namely 5 KN, 15 KN, and 25 KN with a sintering temperature of 800 ° C. 0.003734 gr / mm3. While the lowest density value is found in the composition 75:25 Wt.%, The compacting force is 5 KN with a value of 0.002436 gr / mm3. The decrease in the density value is influenced by the composition and compaction force applied. The greater the borosilicate composition, the lower the density. Meanwhile, the greater the compacting force applied, the density will increase.

1. Introduction

Bone is an active tissue that is metabolically undergoing continuous growth by two processes, namely bone formation (formation) and absorption (resorption). This process depends on the activity of osteoclasts, osteoblasts, and osteophytes [1]. The need for reconstructive of bone is continuously increasing. Now, over 500,000 bone graft procedures are performed annually. The estimated cost of these procedures approaches US$ 2.5 billion per year. Therefore, there is still a constant need to search for material implant for tissue engineering [2].

To replace and repair bone, a material similar to human bone is needed. Human bones contain a compound called hydroxyapatite. One source of the hydroxyapatite material is beef bone [3]. Hydroxyapatite (HA) has the molecular formula $Ca_{10}(PO_4)_6$ (OH) $_2$ is included in the calcium phosphate compound family. Hydroxyapatite is used because of its excellent biocompatibility with hard tissues [4].

Hydroxyapatite bio ceramics are successfully used as material implants as they are chemically similar with the in organic constituent of biological hard tissue. HAP is also a potential implant material due to its excellent osteon conductive properties [5]. The biocompatibility of hydroxyapatite to the mineral of the bone has made hydroxyapatite a potential in material implant. Application of hydroxyapatite in

[1] burmawi@bunghatta.ac.id.

Content from this work may be used under the terms of the Creative Commons Attribution 3.0 licence. Any further distribution of this work must maintain attribution to the author(s) and the title of the work, journal citation and DOI.
Published under licence by IOP Publishing Ltd

critical bone defects has further evolved in tissue engineering with the using cell, scaffold and various growth molecules [6].
Utilizing human bones, hydroxyapatite compounds need to be formed into a composite by adding other elements, by compaction and sintering processes [3]. This sintering will result in changes in the microstructure, including an increase in density values, grain growth, and shrinkage of composite mass [7]. The sintering temperature will improve the bonds present in the material and its density. This effect can be seen on the microstructure and mechanical properties indicated by the grain size, hardness and modulus which will increase with the higher temperature [8]. Density is the value of bone density. Bone density will be affected by the composition, compacting and sintering temperature during manufacture. This factor affects bone density. Bone density ultimately affects bone strength. Bone that has high density will experience fewer fractures and vice versa [9].
This study will examine the density value of the hydroxyapatite-borosilicate composite as a candidate for implant material.

2. Methodology

This research was conducted by mixing the bovine bone hydroxyapatite material with borosilicate material as a composite constituent. The mixing process is carried out using ball milling with the number of balls of 25 pieces with a ball diameter of 10 mm with a rotating speed of 200 rpm for 60 minutes. The mixed bovine-borosilicate hydroxyapatite powder was formed according to the composition by being compacted with a compressive force of 5 KN, 15 KN and 25 KN. In order for the fusion process to occur and the emergence of a mechanical bond in the composite, the sintering process was carried out at a temperature of 800°C for 3 hours.

3. Result and Discussion

After measuring the density, the density value of Hydroxyapatite-Borosilicate was obtained with a compaction force of 5 KN, 15 KN and 25 KN with a temperature of 800 ° C.

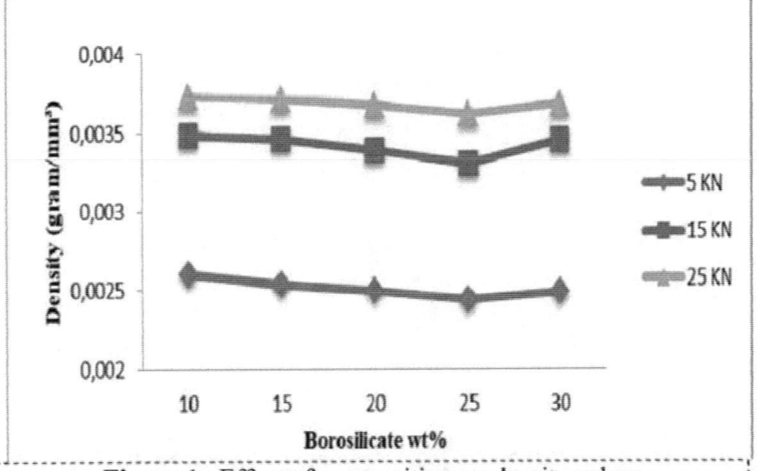

Figure 1. Effect of composition on density values

Figure 1 shows the relationship between composition and density values, starting from the composition of 90:10% wt. with a density value of 0,003732 gr/mm3, then decreasing the density value until the composition of 75 :25 wt.% 0,002436 gr/mm3. The decrease in density in influenced by the density size of each powder. From the material data, it is known that hydroxyapatite powder has a greater density than borosilicate powder. So that the mixing of hydroxyapatite with borosilicate which is done by ball milling process causes changes in density.The more borosilicate compositions were shown to reduce the density of the hydroxyapatite-borosilicate composites

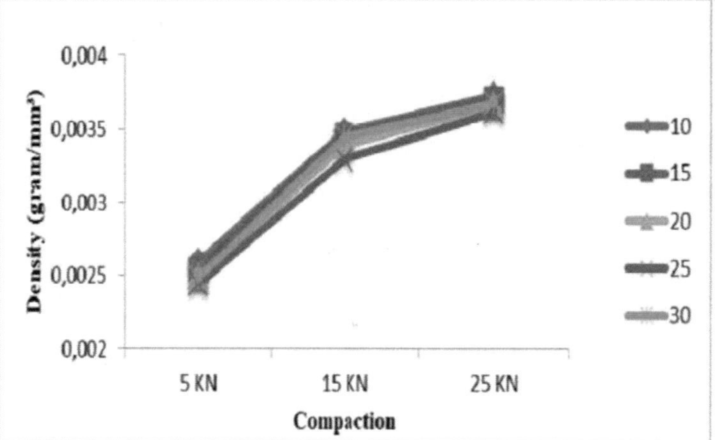

Figure 2. Effect of composition on density values

Figure 2. shows the relationship between the compacting force and the density value, from the data obtained it shows that the higher the compacting force, the higher the density value, this value can be seen in the compacting force of 25 KN with a composition of 90:10% by weight.

The higher the compacting force, the greater the pressure received by the powder to become denser the hydroxyapatite-borosilicate mixture so that it increases the density of the composite material. With the presence of a sintering temperature as high as 800 ° C given to the Hydroxyapatite-borosilicate composite this resulted in a diffusion process. When the diffusion process occurs between hydroxyapatite atoms and borosilicate atoms, a mechanical bond will arise between hydroxyapatite powder and borosilicate powder. The combination and mechanical bonding that occurs is influenced by the grain size of the hydroxyapatite which is smaller than the borosilicate, so the more the hydroxyapatite composition, the higher the density. Likewise with the compacting force, the higher the compacting force also causes the density to also increase.

4. Conclusion

The highest density value is found in the compacting force of 25 KN with a composition of 90:10 Wt.% With a density value of 0,003732gr / mm3. While the lowest density value is found in the 5 KN compacting force with a composition of 75:25 wt. % of 0,002436gr / mm3

References

[1] Seibel MJ. Biochemical markers of bone turnover part I: biochemistry and variability. ClinBiochem Rev 2005; 26: 97-122.

[2] Puska M, Allan J, Aho and Pekka V, 2012, Polymer composites for bone Reconstruction, Advances in Composite Material - Analysis of natural and Man-made material, www.intechopen .com

[3] Burmawi, Novesar J, SyukriArif, Gunawarman, Characterization hydroxyapatite from bovine bone with Mechanical Technic, Oriental Journal Chemistry, 2018.

[4] Kusrini, E., Sontang, M. 2012. Characterization of X-Ray Diffraction and Electron Spin Response: Effects of Sintering Time and Temperature on Bovine Hydroxyapatite. Rad. Physical and Chem. 81, hal 18-125.

[5] Singh, A, 2012, Hydroxyapatite, a biomaterial: Its chemical synthesis, characterization and study of biocompatibility from shell of garden snail, helix aspers, Bull material Science, 35, 6, pp 1031-1038.

[6] Krishnamurithy, 2013, A review on hydroxyapatite-based scaffold as a potential bone graft substitute for bone tissue engineering application, JUMME, 2013, 16 (2)

[7] Dorre and Hunber. 1984. Alumina Library of congress cataloguing in publication Data. Cambridge. USA

[8] Heidi, Lynn P, Hysitron, Application note, University of Wisconsin-Madison , Amanda Simpson, Hysitroninc
[9] Hallman, M & Thor A, 2000, Bone substitute and growth factor as an alternative / complement to autogenous bone for grafting in implant dentistry, Journal of Compilation Periodontology 2000, 47, 2008, 172-192.

IOP Conf. Series: Materials Science and Engineering **990** (2020) 012025 doi:10.1088/1757-899X/990/1/012025

The influence of Shot Penning on fatigue crack growth rate of Chemical Milling product Al-2524-T3 alloys which has been Stretched

Yovial[1], Jamasri[2], Yos Nofendri[3], W Marthiana[1], Duskiardi[1]
[1]Mechanical Engineering Department, Universitas Bung Hatta, Jl. Gajah Mada no 18, Padang, Indonesia.
[2]Mechanical Engineering Department, Universitas Gajah Mada, YogyakartaIndonesia.
[3]Mechanical Engineering Department, Universitas Muhammadiyah Prof. Dr. Hamka, Jakarta, Indonesia.

E-mail: yovial@bunghatta.ac.id

Abstract. This study aims to investigate the influence of shot penning on fatigue crack growth rate of chemical milling product Al 2524-T3 which have been stretched. This alloys were stretched beyond yield stress, namely 1%, 3% and 5% of each, and then performed chemical milling process of one side. The etching used in chemical milling process were $NaOH+Na_2S+H_2O$ solutions with certain concentration. The surface was performed shot penning process with varying intensity of 0.03 A, 0.05 A and 0.07 A, respectively. These material were then tested its crack growth rate. The crack growth rate tests were performed under 20% stress level, stress ratio of 0.3 and at frequency 7-13 Hz. The average of crack propagation data against the cycle number were processed using an incremental method, in accordance with ASTM E647 standard, to obtain crack growth rate curve (da/dN vs. ΔK). The results show that the increase of shot penning intensity has no significant effect on the fatigue life, with the growth average of 4,5%. The highest increase occurred at the shot intensity of 0.03 A, 0.2 times its chemical milling, while the lowest decrease occurred at the shot intensity of 0.07A, 0.012 times its chemical milling. The crack retardation effect at the beginning of the crack plays a role in reducing the crack propagation rate.

1. Introduction

Sheet metal is widely used in the aircraft and automotive industries, especially for relatively wide components with light weights. Sheet metal is formed using several processes, to produce the desired changes, from simple to more complex shapes [1][2].

In general, sheet metal components on aircraft due to their thin and wide shape are worked with a chemical milling process, which is a non-traditional abrasive process that uses chemical solutions as a medium since working with ordinary machining processes is not supported [3]

1 yovial@bunghatta.ac.id

Content from this work may be used under the terms of the Creative Commons Attribution 3.0 licence. Any further distribution of this work must maintain attribution to the author(s) and the title of the work, journal citation and DOI.
Published under licence by IOP Publishing Ltd

Materials worked by using the chemical milling processes, generally solid aluminum and magnesium, will experience distortion due to the uneven distribution of residual stress due to the release of residual stress during the working process. Its can cause damage to components, due to stress instability when assembling these components [4]

The chemical milling process also tends to produce micro cracks on the surface. Micro cracks in the material will propagate if the material loaded a dynamic work load cycle until it reaches its critical crack length, causing complete damage known as fatigue failure [5]

The residual stress on the material cannot be avoided since it occurs as a result of the sheet metal manufacturing process. One of the residual stresses of the material is due to the initial cold forming process of the sheet metal. The residual stress that arises is not always beneficial in the further forming process. On the other hand, in the further forming process, another problem that arises is the appearance of micro-cracks in the material. But the reduction of residual stress in the material can be done with certain processing processes, one of which is the stretching process with a certain percentage until plastic deformation occurs on the material. The stretching process is expected to reduce the distortion that occurs as little as possible after the material subjected to a chemical milling process.

On the other hand, chemical milling process is often carried out only on one side of the surface of the material, so that the uneven distribution of residual stress creates tensile stress on one side of the process and compressive stress on the other. A treatment that can have a positive effect on such stress forms is to apply compressive stress to the surface of the material by using techniques, such as shot penning or surface rolling. The surface layer of the material, in the shot penning process, will undergo plastic deformation down to a depth, resulting in a compressive residual stress.

This study aims to determine the crack propagation of the Al 2424-T3 stretching products which undergo a chemical milling process which is then shot penning. This study waspart of the continuation researchconducted by Yovial *et al.* [6].

2. Methodology

2.1 Material

The materials used in this study are aluminum alloy types Al 2524-T3 in sheet form, with the following chemical composition: This study is

Table 1. chemical composition of aluminum alloy Al 2524-T3

Material	Chemical composition (wt%)						
	Cu	Mg	Mn	Fe	Si	Zn	Ti
Al 2524-T3	1.0	0.03	0.05	0.1	0.3	0.3	0.03

2.2 Procedures

Types of specimens for testing, consist of a) raw material, b) specimen from the result of stretching+chemical milling processes, with a stretching of 1%, 3% and 5%, c) specimens from the result of stretching+chemical milling+shot penning processes, with shot intensity penning 0.03 A, 0.05 A and 0.07 A.

The stretching process is the first treatment applied to the material before it is given the chemical milling and shot penningprocesses. The percentage of stretching refers to the percentage of the yield stress of the material (σ_{ys}). The stretching method used is pulling material using a stress that is several percent higher than the yield stress. Therefore:

$$\sigma_{1\%} = \sigma_{ys} + 1\% x \sigma_{ys} \tag{1}$$

The stress at which plastic deformation began observable is taken using the offset yield strength, at a strain of 0.2 percent. Furthermore, the value of the stress is used as a benchmark for measuring the next stretching stress, as given by equation 1.

The chemical milling process is carried out in stages, according to established procedures. Sheet metal soaked in a bath containing $NaOH + Na_2S + H_2O$ solution with certain concentrations. The duration of soaked depends on the desired abrasive thickness in the material.

The specimens in this study followed the ASTM E-647 standard, as shown in Figure 1. The initial thickness of the specimens was 1.6 mm. Chemical milling process removed the center of the surface of the test area by a thickness of 0.2 mm, while the thickness of the two sides which was applied with the masking did not change. Masking is the process of applying the masking material to the surface to ensure that only desired areas were etched. The initial crack, $2a_0$, in the test specimen was made 10 mm long and located in the middle of the specimen by using an EDM machine. The súrface of the test plane was then shot penning with varying intensity, namely 0.003 A, 0.005 and 0.007 A.

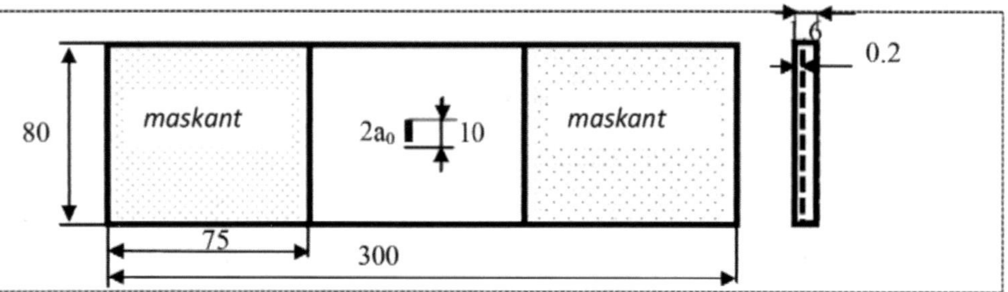

Figure 1. ASTM E-647 specimen

2.3 The seven-point incremental polynomial method

The fatigue crack propagation test carried out by a servo pulsar machine that operates at a frequency of 7-15 Hz. The crack length was measured using a traveling optical microscope from each side of the specimen surface, namely the left side of the crack on the first surface and the right side of the crack on the second surface. The optical traveling microscope has a magnification of 20x and the scale for each strip is 0.1 mm. Testing using a stress level of $S_L \pm 20\%$, was carried out at a stress ratio of R = 0.3.

The maximum loadgiven, P_{max}, calculated by the equation:

$$P_{max} = \frac{S_L \sigma_{ys} t w}{1000} \text{ kN} \tag{2}$$

where:

σ_{ys} = yield stress of raw material (kg/mm^2), t = thickness (mm)
S_L= stress level (%), w= width (mm)

Meanwhile, the minimum load obtained from:

$$P_{min} = R.P_{max} \text{ kN} \tag{3}$$

The crack propagation data obtained were in the form of crack length (a_i) and cycle (N_i). The average crack length at a given cycle were

$$a = \frac{(a_{kiri} + a_{kanan} + 2a_o)}{2} \tag{4}$$

The scattered data processed by the 7-point incremental polynomial method[7], which provides the equation for a collection of local points in the form of:

$$a_1 = b_o + b_1\left(\frac{N_i - C_1}{C_2}\right) + b_2\left(\frac{N_i - C_1}{C_2}\right)^2 \tag{5}$$

for

$$-1 \leq \left(\frac{N_i - C_1}{C_2}\right) \leq 1 \tag{6}$$

Constants b_0, b_1 and b_2 are the regression parameters obtained from the least square method in the range of $a_{1-n} \leq a \leq a_{1+n}$, while, the parameter $C_1$and C_2were calculated by the equation:

$$C_1 = \frac{1}{2}(N_{i-m} + N_{i+m}) \tag{7}$$

$$C_2 = \frac{1}{2}(N_{i+m} - N_{i-m}) \tag{8}$$

The crack propagation rate at the N_i cycle value and the crack length a_i obtained from the equation:

$$\frac{da}{dN} = \frac{b_1}{C_2} + 2b_2\left(\frac{N_i - C_1}{C_2{}^2}\right) \tag{9}$$

The stress intensity factor (ΔK) at the number of $N_i$cycles calculated from [8]:

$$\Delta K = \frac{\Delta P}{B}\sqrt{\left(\frac{\pi a}{2w} sec\frac{\pi a}{2}\right)} \tag{10}$$

Where

α = 2a/w (mm)
B = thickness (mm)
$\Delta P = P_{max} - P_{min}$(kg)
W= width (mm)

3. Result and Discussion

Some of the crack propagation results test are shown in Table 2, with the number of fracture cycles of the specimen reached 108 cycles.

Table 2. The example result of the crack propagation test of Al 2524-T for 1% + CM + 0.03A

No	a_{left} (mm)	a_{right} (mm)	N (cycles)	a_{av} (mm)	da/dN (m/cycles)	ΔK ($MPa.m^{1/2}$)
1	5,1	4,95	0	5,025		
2	5,1	5,1	5090	5,1		
3	5,2	5,2	5610	5,2		
4	5,2	5,3	6150	5,25	1,27E-07	11,8278327
5	5,3	5,4	6820	5,35	1,39E-07	11,9469025
6	5,3	5,5	7370	5,4	1,33E-07	12,03533074
7	5,4	5,6	7830	5,5	1,32E-07	12,11396239
			....			
102	23,2	23	37256	23,1	6,81E-06	31,04118591
103	23,8	23,5	37353	23,65	9,03E-06	32,09115657
104	24,4	24	37413	24,2	1,20E-05	32,92689248
105	25	24,5	37455	24,75	1,70E-05	33,7825963
106	25,3	25	37476	25,15		
107	26	25,5	37512	25,75		
108	27	26	37528	26,5		

3.1 Stretching effects

Table 2, were than plotted in a vs. N, and combined with the test results of several specimens with different processes. This graph, as in Figure 2.1, showsthat the higher the stretching percentage, the lower the number of testing cycles.

In the form of da/dN vs. ΔK curve, a clear difference is shown between the stretching percentages of 3% and 5%. Figure 2.2 shows the da/dN vs. ΔK curve appears to be shifted more to the left at the percentage of 5%. The curve shift indicates a decrease in fatigue life. In other words, the more to the left of the da/dN vs. ΔK curve, the lower the fatigue life of the material.

This phenomenon indicates that the ductility of the material decreases with the higher the stretching percentage. Ductility is related to the toughness of the material, namely the ability of the material to absorb energy in the plastic area. Therefore, the higher the stretching percentage, the less tough the material will be. The decrease in toughness is also caused by the reduction in residual stress due to the chemical milling process given earlier, so that the residual stress cannot accommodate the given dynamic tensile stress.The chemical milling process causes a reduction in residual stress on the material, especially on the surface, due to the erosion. This reduction causes an imbalance of residual stress distribution which is compensated by the material in the form of distortion. After the material is stretched, the residual stress it has is reduced, so the material no longer experiences distortion [9].

It can be concluded from the estimated life cycle, the increase in the stretching percentage from 1% to 3% reduces the fatigue life by 0.0776 times, and more decreased by 0.107 times at the 5% stretching percentage.This phenomenon also shows that the increase of stretching percentage of aluminium alloy Al 2524-T3 does not appear to be effective in increasing the fatigue life of the material

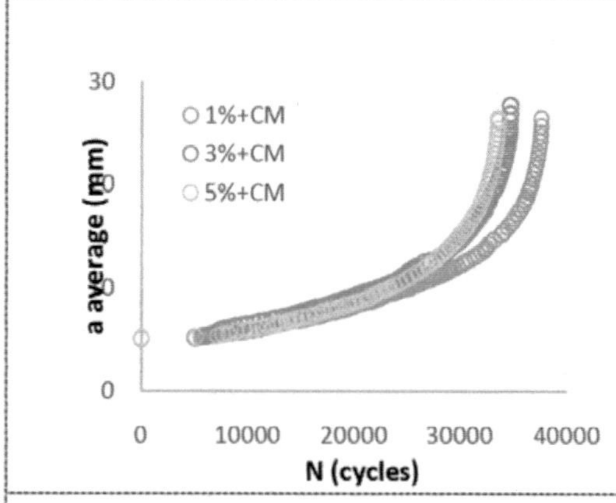

Figure 2.1. a vs N graph

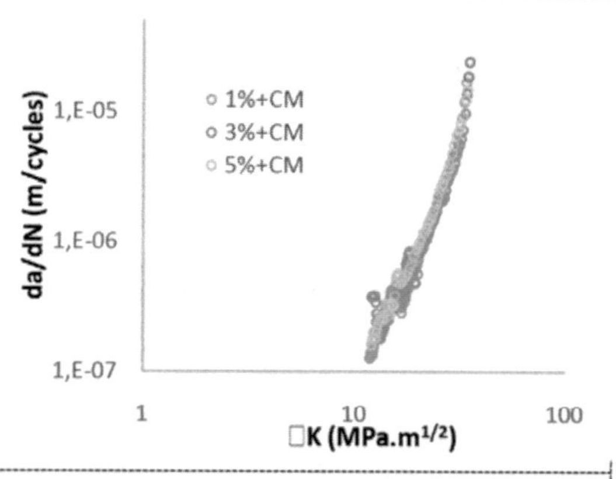

Figure 2.2. da/dN vs ΔK graph

3.2 Shot Penningeffects

Figure 3.1shows an a vs. Ngraph of the process with different shot intensities. It also shows that the higher the intensity of shot penning, the lower the number of testing cycles. The same phenomenon is also found in other materials with different processes (different in stretching percentage), which are not shown in this paper. However, for a low shot intensity, the increase in the test cycle that occurs is quite significant, and then decreases as the shot intensity increases. This phenomenon indicates that the chemical milling product which is given a shot penning process with low intensity significantly increases the number of test cycles.

In the form of da/dN vs. ΔK curve, as shown in figure 3.2, 3,3 and 3,4 respectively,a clear difference is shown between the shot intensities of 0,03 A, 0,05 A and 0,07 A.Figure 3.2 shows the da/dN vs. ΔK curve appears to be shifted more to the right at the shot intensities of 0,03 A. The curve shift indicates an increase in fatigue life. In other words, the more to the right of the da/dN vs. ΔK curve, the higher the fatigue life of the material.

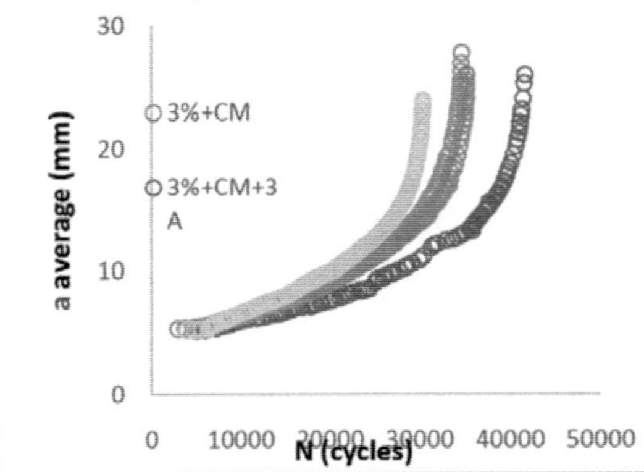

Figure 3.1. a vs N graph

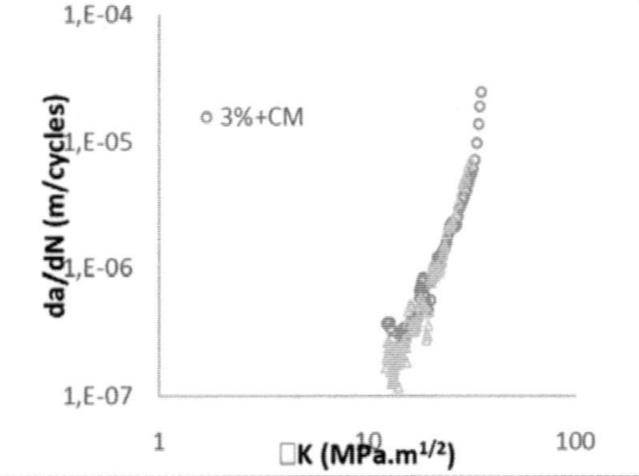

Figure 3.2. da/dN vs ΔK graph

The shot penning intensity is related to the impact or the number of impacts per surface area. The direct result of this is the occurrence of deformations at a certain depth. The greater the intensity of the shot, the greater the impact energy received by the material, so that the more plastic deformation surface area will be.The deformation that occurs will cause the dislocation density to increase, which in turn results in a buildup of dislocations in the slip plane. The buildup causes a high dislocation density, so that the movement of the dislocations will inhibit each other; in other words, a stuck dislocation causes the material to become hard and strong. In other word, shot penning should increase the resistance to the development of crack tip plasticity, or have an impact on the arrest of fatigue cracks [10].

It can be concluded from the estimated life cycle, an increase in shot intensity of 0.03 A increased fatigue life by 0.2 times, although it still increased at a shot intensity of 0.05 A, but it had decreased to 0.016 times. However, the life cycle reduced by 0.032 times at the shot intensity of 0.07 A.

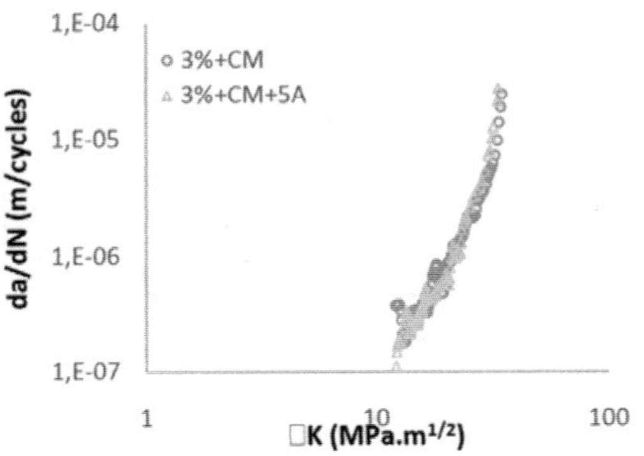

Figure 3.3da vs N graph

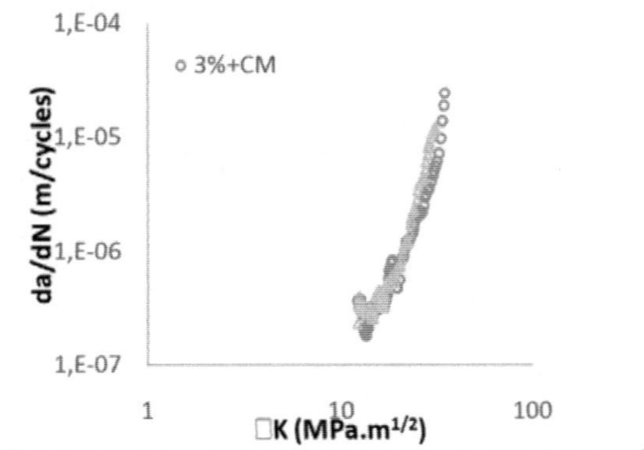

Figure 3.4da/dN vs ΔK graph

4. Conclusion

Sheets of aluminium Al 2524-T3 alloy wereprepared by a different process that involved stretching, chemical milling and shot penning. The results shown that effect of stretching percentage reduces the

IOP Conf. Series: Materials Science and Engineering **990** (2020) 012025 doi:10.1088/1757-899X/990/1/012025

fatigue life of material. The higher the stretching percentage, thelower thefatigue life. On the other hand, shot intensity increased the fatigue life of material to certain intensity.

Reference

[1] L H Vinod and R SShivashankar 2018 *materials today:proceedings*, **5**(1): 2564-2574
[2] M. Gottschalk and G. Hirt 2016*AIP Conference Proceeding*, **1769**(1)
[3] D S Philippe, G Germain, S Boude and E Giroud2013*Int JComp Integr Manufac*, **27**(2).
[4] N Rui, D Teresa, Jorge, L Alves, F Torres, J Pinto 2017 *Ciência & Tec Mat*, **29**(1), e40-e45
[5] S Birhan, R Gaddam, J J. Roa, M Antonio, M Antti, R Pederson 2016*Int J Fatigue*, **92**(1), 193-202.
[6] Yovial, Jamasri, Wenny, Rizky and Suryadimal 2020*Jurnal Kajian Teknik Mesin* **5**(1) 36-41
[7] J Zheng and BEPowell 2008 *Fatigue & Fracture Eng Mat & Structures,***20**(9), 1341–1350
[8] JR Mohanty, B B Verma and PK Ray 2010*Int. J Microstructure and Mat Prop*, **5**(1), 79-87
[9] CF Yao, Xing TD, D. Wu, Zheng Z, JY Zhang 2016, Advances Mech Eng, **8**(10) 1-10
[10] Bert L. Smith,TanyaLZ, Floresand Ala L Hijazi 2005 J Aircraft **42**(2)

IOP Conf. Series: Materials Science and Engineering **990** (2020) 012026 doi:10.1088/1757-899X/990/1/012026

Performance Experimental Study on Stepped Type Basin Solar Still To Produce Salt and Fresh Water

Mulyanef[1],Kaidir[1], Kadafi Z [1], K Sopian[2]
[1]Department of Mechanical Engineering, Universitas Bung Hatta, 25243 Padang, Indonesia
[2] Solar Energy Research Institute, UniversitiKebangsaan Malaysia, 43600 Bangi, Selangor Malaysia

E-mail: mulyanef@bunghatta.ac.id

Abstract. This paper presents an experimental performance study of a stepped type solar still for processing seawater into salt and fresh water. Its performance was tested in Padang City, under Indonesian environmental conditions. The experimental setup this study consisted of a tank of seawater, solar distillation and distilled water storage tanks. The working principle is the distillation of sea water put into the basin to be heated with solar power. The water will evaporate and condense on the surface of the cover glass distillation, distillation of fresh water stored in the container. Warming in the distillation carried out continuously until the water evaporates and the depleted salt crystals formed in the basin. The result shows that the productivity of the solar still was highly affected by the increase in solar radiation and ambient temperature. The productivity was 176 gr salt and 3,662 ml water with the average solar intensity 566 W/m². The time it takes to produce salt is fourdays.

1. Introduction

In daily life, salt and fresh water are very important needs for the human body. In several areas in Indonesia, the needs of salt and fresh water cannot fullfilled yet by the government. Indonesia's domestic salt production is only 31%, so to meet the need for salt, it still has to be imported from outside Indonesia. This is very unnatural considering that Indonesia is a country that has beaches stretching from Aceh to Papua. One of the efforts to provide salt and fresh water is solar still. The main advantage of solar stills is that solar energy is freely available, free of charge and does not utilize fossil fuels. Many studies have been carried out for the purpose of increasing the productivity of basin type solar stills, among others; [1]. discussed the role of climatic, design and operational parameters on the productivity of solar still – basin type, and they concluded that the productivity of solar still–basin type requires more research and development to optimize the absorption, evaporation and condensation processes. [2]. investigated the installation of an internal reflecting mirror in a conventional single slope solar still. The area of the solar still basin is 80 cm × 80 cm. The productivity of solar stills with and without mirrors is 2,077 liters and 1,408 liters. The effect of the tilt angle on productivity and the optimum value of the glass cover tilt angle was reviewed [3].Conclusion shows that the angle of inclination of the glass cover should be large in winter and small in summer.[4] studied the effect of heat energy storage for transient performance of stepped solar still.

[1] mulyanef@bunghatta.ac.id

Content from this work may be used under the terms of the Creative Commons Attribution 3.0 licence. Any further distribution of this work must maintain attribution to the author(s) and the title of the work, journal citation and DOI.
Published under licence by IOP Publishing Ltd

IOP Conf. Series: Materials Science and Engineering **990** (2020) 012026 doi:10.1088/1757-899X/990/1/012026

The results showed that solar still modification is efficient for producing water in times of lack of sunlight at night. [5]. studied the effects of variations in both salt water depth and basin width on the performance of stepped solar, theoretically and experimentally. The results show that, the maximum productivity of stepped solar stills is approximately 57.3% higher than conventional solar stills in a basin depth of 5 mm and a width of 120 mm. [6]. conducted research on solar stills with reflectors for producing salt and fresh water.The results showed that there was a 16.8% increase in fresh water production in collectors using reflectors compared to collectors without reflectors. The time to produce salt is four days with an average solar intensity of 696 W / m2. Then, [7]. presents a modified stepped solar still via internal and external reflectors. The results show that the productivity of a modified stepped solar still with internal and external reflectors is 125% higher. than conventional solar still. [8] conducted an experimental study of stepped solar still integrated with reflectors and external condenser. The result is that the productivity of stepped solar stills using a condenser is higher than conventional solar stills. In this study performance experimental on stepped type basin solar still to produce salt and fresh water were investigated.

2. Experimental Setup

Fig. 1 shows a schematic diagram of a stepped solar still. The area of the solar still absorber plate is 0.72 m^2. Solar still is covered with commercial clear glass with a thickness of 5 mm and an angle of 30 degrees of the glass cover. The stepped solar still has the same construction and geometry, the absorbent plate is made of 3 sidesteps. The lower wall of the absorber plate is insulated with 30 mm polyurethane foam (thermal conductivity 0.021 W/m K). The working principle of the stepped solar still is that seawater flows from the water tank into a multilevel basin. Solar radiation shines on the absorbent plate and turns it into heat. Sea water in the basin becomes hot and evaporates and then, water vapour collects under the cover glass. Condensation occurs on the inner covering glass due to the difference in temperature inside the collector with the temperature outside. Then the condensed water flows through the tube to the water container. Testing is carried out continuously until the sea water in the basin evaporates to produce salt crystals.

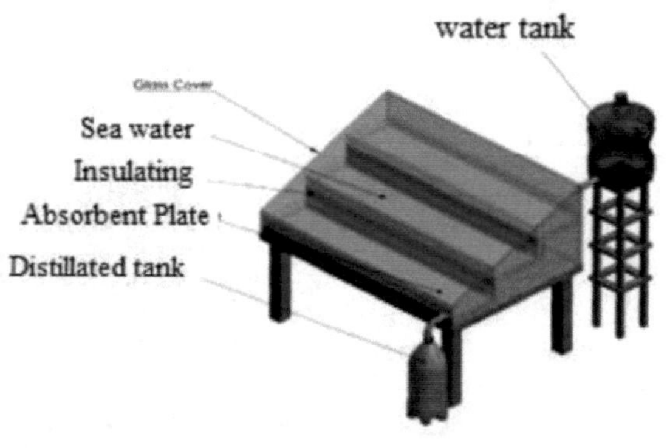

Figure. 1. Schematic Diagram of Stepped Solar Still

3. Results and Discussion

In this section, the performance of basin type stepped solar still is verified under different evaluation parameters. The parameters are the temperature of the seawater in the basin, the temperature of the cover glass, the temperature of the absorber plate, the temperature difference between sea water and the glass cover.

3.1. Effect of solar radiation on the performance of the solar still

Fig.2 shows the relationship between temperature and time on the first day of the experiment. The graph shows variations in solar radiation, ambient temperature (Ta), absorbent plate temperature (Tp), evaporation temperature (Tsv), sea water temperature (Tw) and glass cover temperature (Tg). Its value increases over time until it reaches its maximum value at midday. Furthermore, the high seawater temperature in the basin causes an increase in evaporation and condensation rate, thus an increase in the temperature of the still stepped solar glass.

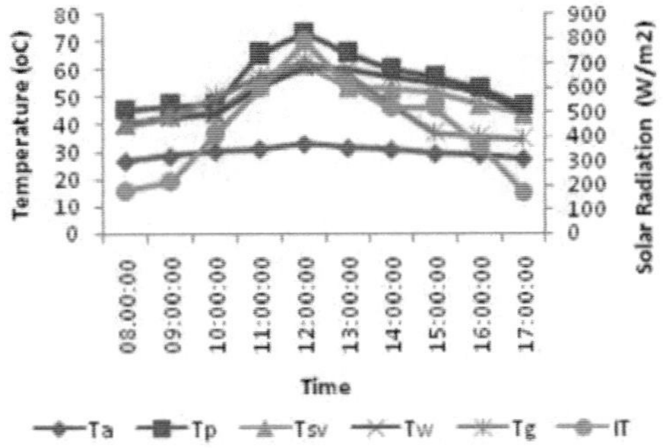

Figure 2. Relationship Between Temperature, Time and Solar Radiation

3.2. Salt and Fresh Water Productivity

Figure 3. shows the relationship between the value of solar radiation intensity and freshwater productivity. In the graph, it can be seen that in the morning the intensity of solar radiation and freshwater productivity is still low, it continues to increase until it reaches the maximum value during the day. Then the intensity of solar radiation and fresh water productivity decreases in the afternoon. The higher the intensity of solar radiation, the higherthe productivity of the fresh water produced. This is because the low sea water temperature in the basin in the morning takes longer to warm up. In addition, it can be seen that the maximum productivity occurs at the maximum sea water temperature.

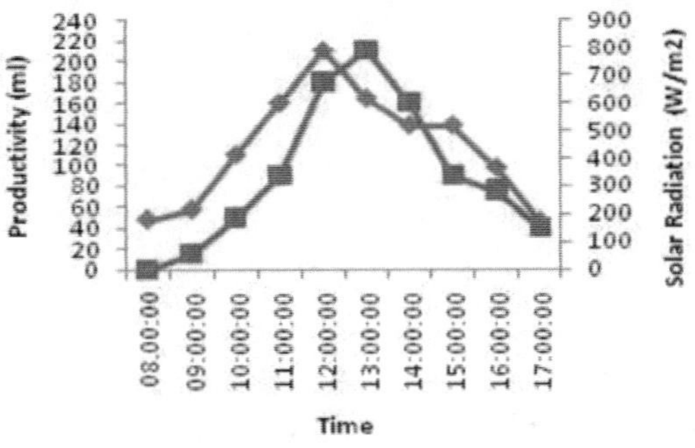

Figure 3. Relationship Between Solar Radiation Intensity and Freshwater Productivity

Figure 4 shows a graph of freshwater productivity and average solar radiation intensity with sea water volume varies from stepped solar still. The test was carried out by varying the volume of sea water in the basin (6,000 ml, 4,500 ml and 3,000 ml). The test results showed that the highest freshwater productivity wasproduced in the volume of sea water in the basin of 6,000 ml, while the lowest was

IOP Conf. Series: Materials Science and Engineering **990** (2020) 012026 doi:10.1088/1757-899X/990/1/012026

produced in the basin volume of 3,000 ml. This is because the more sea water in the basin the more fresh water productivity is produced and the longer it takes to produce salt.

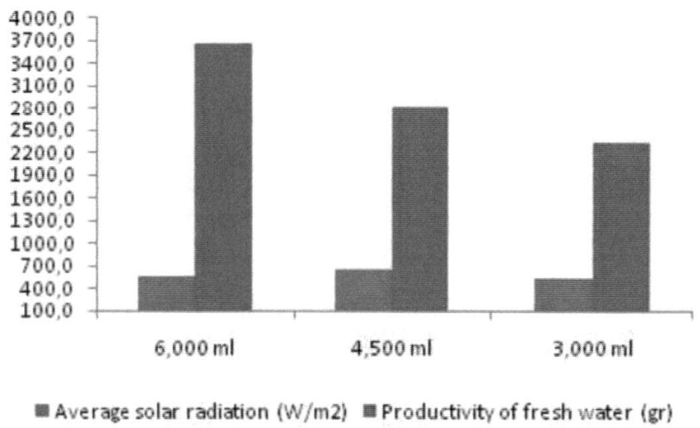

Figure 4. Freshwater Productivity and Average Solar Radiation Intensity With Different Sea Water Volume

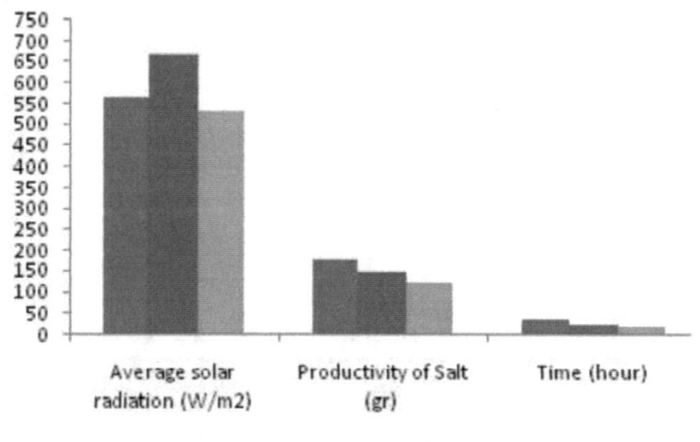

Figure 5. Average Solar Radiation Intensity, Salt Productivity andTime and Volume of Sea Water in the various Basin

Figure 5 shows a graph of the solar radiation intensity and salt productivity of a stepped solar still. The test was carried out by varying the volume of sea water in the basin (6,000 ml, 4,500 ml and 3,000 ml). The volume of sea water in the basin is 6,000 ml and the average solar radiation intensity is 566 W / m², it takes 36 hours to produce salt. The volume of sea water in the basin is 4,500 ml and an average solar radiation intensity of 667 W / m² takes 22.5 hours to produce salt (176.5 gr). The volume of sea water in the basin is 4,500 ml and an average solar radiation intensity of 531 W / m² takes 18 hours to produce salt (146.5 gr). The productivity of water and salt in stepped solar still depends on the volume of sea water in the basin and the average intensity of solar radiation. The higher the intensity of solar radiation, the higher the productivity of fresh water and accelerate the production of salt (118.9 gr). Similar results were previously produced by the previous [6, 8].

IOP Conf. Series: Materials Science and Engineering **990** (2020) 012026 doi:10.1088/1757-899X/990/1/012026

4. Conclusions

Based on the results of tests carried out on stepped solar still, it can be concluded as follows:

1. The higher performance of stepped solar still is achieved at 6,000 ml deep sea water volume, resulting in 176.5 g of salt and 3,662 ml of fresh water. It takes 4 days to obtain the salt, with an average solar intensity of 566 W / m².
2. The productivity of freshwater stepped solar still decreases with increasing sea water depth in the basin.
3. The value of sun intensity affects the temperature of the absorbent plate, where the higher the intensity of the sun, the higher the temperature of the absorbent plate.
4. The value of the sun's intensity affects the time to produce fresh and salt water, where the higher the sun's intensity, the shorter the time it takes to produce salt.

Acknowledgment:

I would like to thank the Rector of Universitas Bung Hatta who has funded the 2020 Research Program No. 068-08/LPPM-P/Hatta/IX-2020. Also, I would like to thank LPPM for providing resources, so that work can run smoothly.

Reference:

[1] Ali.F. Muftah, M.A. Alghoul, Ahmad Fudholi, M.M. Abdul-Majeed, K. Sopian, Factors affecting basin type solar still productivity: a detailed review, *Renew. Sust.Energ. Rev.* **32** (2014) 430–447.

[2] Abdallah S, O. Badran, M. Abu-Khader, Performance evaluation of a modified design of a single slope solar still, *Desalination* **219** (1–3) (2008) 222–230.

[3] Khalifa AJN. On the effect of cover tilt angle of the simple solar still on its productivity in different seasons and latitudes. *Energy Convers Manage* 2011;**52**:431–6.

[4] Radhwan AM. Transient performance of a stepped solar still with built-in latent heat thermal energy storage. *Desalination* 2004; **171**:61–76.

[5]Kabeel AE, Khalil A, Omara ZM, Younes MM. Theoretical and experimental parametric study of modified stepped solar still. *Desalination* 2012; (**289**):12–20.

[6] Mulyanef, Duskiardi, K Sopian, Kaidir, and Zulfika Rahman. Performance Experimental Study of Solar Still With Reflector To Produce Fresh Water and Salt, *Matec Web of Conferences* **248**, 100, 4-8

[7] Omara ZM, Kabeel AE, Younes MM. Enhancing the stepped solar still performance using internal and external reflectors. *Energy Convers Manage* 2014; **78**:876–81

[8] Y. A. F. El-Samadonya, A. S. Abdullah & Z. M. Omara. Experimental Study of Stepped Solar Still Integrated with Reflectors and External Condenser. *Exp. Heat Transf.*, vol. **28**: 392–404, 2015.

IOP Conf. Series: Materials Science and Engineering **990** (2020) 012027 doi:10.1088/1757-899X/990/1/012027

Performance Study ff Increasing Power Plant Efficiency by Reducing Condenser Pressure in Teluksirih Power Plant

Kaidir[1],Burmawi[1] , Mulyanef [1], Muhammad DG [1]
Mechanical Engineering, Universitas Bung Hatta, Padang, Indonesia

E-mail: irkaidir@bunghatta.ac.id

Abstract. Researchonincreasingtheefficiencyof powerplantneedstobedonebecause thecostof producing electricity isquitelarge.Thebiggest costoftheelectricityproductionprocessisthecost offuel.Onewaytoreducefuelconsumptionisto reducethepressure ofthecondenser. By decreasing thecondenser pressure,thesteam turbineloadwillincrease,andtheefficiencyofthe powerplantincreases.Therefore,itisnecessaryto conduct research toanalysispower plant efficiency, analysis therelationship between condenserpressureand powerplantefficiency, determinecoalusagerequirements. Theresulting efficiency willberelatedtotheenergygenerated. Thehighertheenergy generated;theefficiency willalsoincrease.

1. Introduction

Most of the electricity produced throughout the world today is from steampower plants and improving the performance of power plants is crucial to minimize thegreenhouse gas emissions and fuel consumption. Energy efficiency of a thermal powerplant strongly depends on its boiler-condenser operating conditions. The low-pressure endconditions of a condenser have influence on the power output, steam consumption andefficiency of a plant [1]. Electricity isaveryimportant requirementineveryday life.Weusually useelectricity forsimplethingssuchas turning onlights,watching television, charging cellphones,cooling food, washing andsoon.Butintheprocessof producingelectricity itisnotsimple, sometimes itrequirescomplexprocesses and highcosts. The highestcostsintheprocess of electricity production,especially inthe steampower plantareproductionfuels, namely coal.Oneefforttoreduce productioncostsisto minimizetheuseof coalby increasingtheefficiency ofthe power plant [2]. Fig. (1) shows the thermal cycle of the unit boiler combustion system generating scheme at 100% load.One efforttoimprove powerplant efficiency istoreducethepressureofthe condenser.Acondenserisadeviceusedto convertsteamintoa liquidform. Condenser pressureplaysanimportant roleinmaintaining unitoperations [3].Fig. 2. Thermal cycle of the Coal Feed Boiler generating system at 100% full load. The unitcannotoperateproperly whenthe condenserpressureishigh,thisisbecause the steam inLPTurbinecannot be sucked intothe condenser properly.The condenserpressureaffectsthe efficiency oftheturbine,sothatitwillalsoaffectthe efficiency ofthepowerplantandthe production costs [4,5,6,10,11]. Fig. 3. Thermal cycle of the unit main steam system generating scheme at 100% full load.

[1] irkaidir@bunghatta.ac.id.

Content from this work may be used under the terms of the Creative Commons Attribution 3.0 licence. Any further distribution of this work must maintain attribution to the author(s) and the title of the work, journal citation and DOI.
Published under licence by IOP Publishing Ltd

IOP Conf. Series: Materials Science and Engineering **990** (2020) 012027 doi:10.1088/1757-899X/990/1/012027

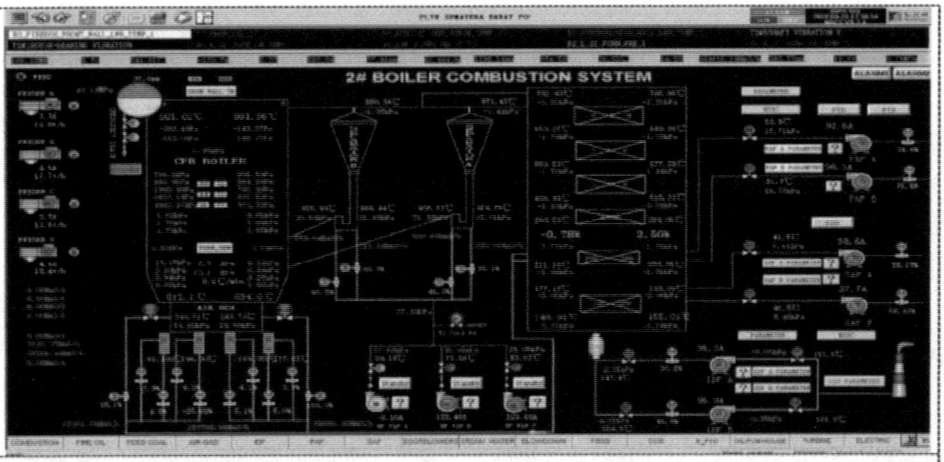

Fig. 1. Thermal cycle of the unit generating scheme boiler combustion system at 100% full load.

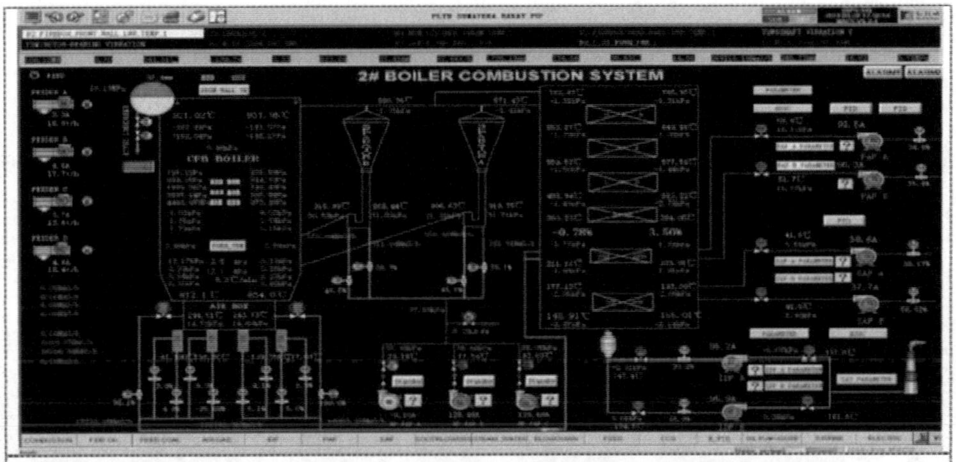

Fig. 2. Thermal cycle of the Coal Feed Boiler generating system at 100% full load.

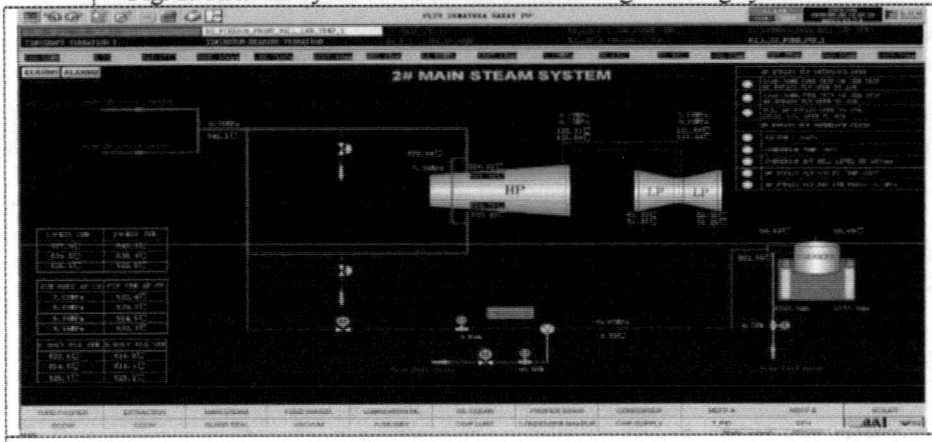

Fig. 3. Thermal cycle of the unit main steam system generating scheme at 100% full load.

IOP Conf. Series: Materials Science and Engineering **990** (2020) 012027 doi:10.1088/1757-899X/990/1/012027

2. Methodology

The researchwascarried out by reducing thecondenserpressureand analysis theimpactontheworkofthe steamturbine.Theeffectof acondenser vacuumonthe powerplantsystemisat leasttwothings.First,increase thesteam turbine load.The vacuumcondenser will affectthe highandlowloadgeneratedby the steamturbine.If thevacuumishigh withtheamountofenergy entering the same turbine itwillgeta higher load. Second, increasing generation efficiency. Theresultingefficiency willberelatedto theenergy generated.Thehigherthe energygenerated,the efficiency willalso increase.

CalculationofHeatRateisobtained from thefollowing equation:

$$NPHR = Q_{in}/NPO$$

Where Q_{in} is heat entering from fuel (Kcal/h). NPO is the network produced (kW)

Calculationofgeneratorefficiency is obtained from thefollowingequation:

$$\eta_g = \frac{859{,}7\frac{kCal}{kWh}}{N\,P\,H\,R}$$

Where859,7Kcal/kWhistheconversion of energyunits.

3. Resultand Discussion

Basedontheresultsofthestudy it wasfoundthatthepower, efficiency, and Heat Rate were as follows:

Table1. Power, EfficiencyandHeatRate

Pressure (bar)	Power (MW)	Efficiency (%)	Heat Rate (kCal/kg)
0.11	93.35	28.57	3009.04
0.10	93.96	28.76	2989.64
0.08	95.44	29.21	2943.25
0.06	97.29	29.77	2887.39
0.04	99.78	30.54	2815.09
0.03	101.10	30.94	2778.55

Table1.showsthatthelower the pressure ofthecondenser,thegreater the powerproduced,the efficiency becomes greater and the heat ratedecreases.

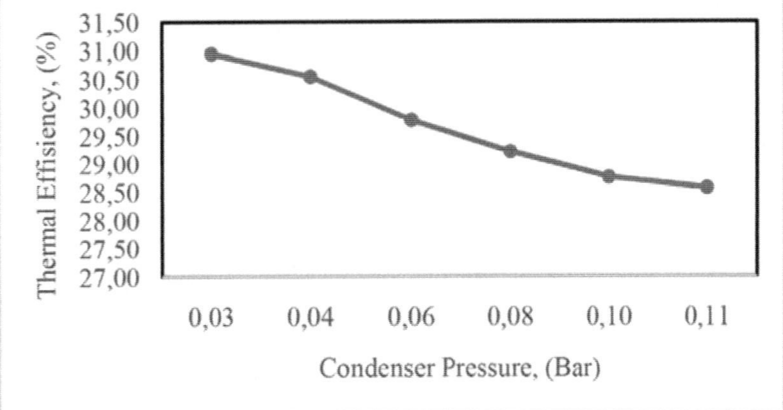

Figure 4. Relationship Between Condenser. Pressureand Power PlantEfficiency

Figure 1. shows the effect of condenser pressure on thermal efficiency. This shows that thermal efficiency decreases with increasing condenser pressure. A decrease in the temperature of the cooling water and so with the pressure of the condenser will cause the output power to rise in the same fuel flow. This also decreases the heat rate of the power plant pressure. Adecreaseinthetemperatureof thecoolingwater andso withthepressure of thecondenser willcause theoutput power torise inthe same fuelflow.This alsodecreasesthe heat rate of the power plant.

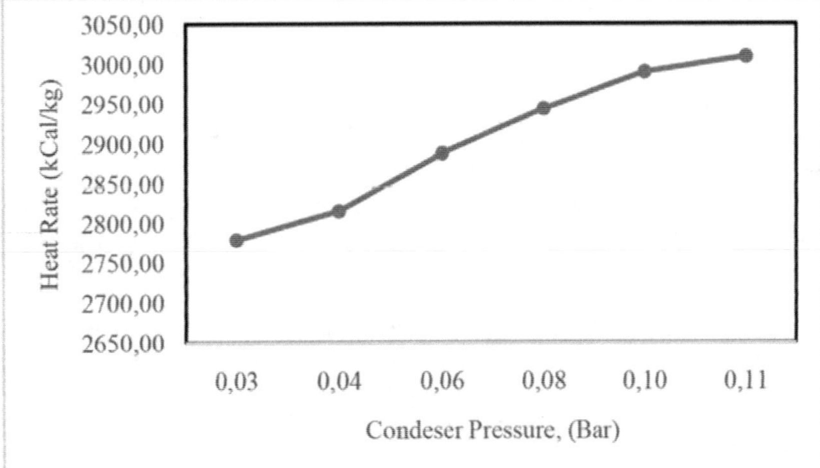

Figure5 .RelationshipBetweenPressureof the Condenserand HeatRate

The moreefficientapowerplantis the lesscoalconsumptionwillbe.This willhaveanimpactonthedecreasingcost of production.Figure 3showsananalysis of the relationshipbetween power plant efficiencytofuelconsumptionintheform of coal.

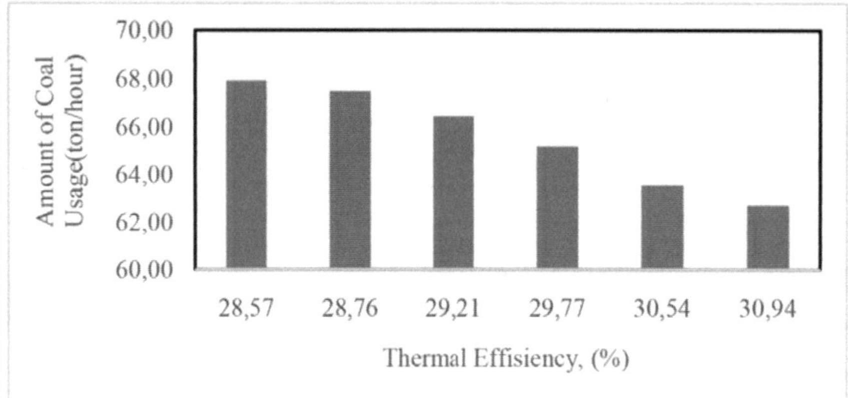

Figure 6. Relationship BetweenGenerator Efficiencyandthe AmountofCoalUsage

4. Conclusion

Basedontheresearchitwasfound that the lower the condenser pressure, the highertheefficiency ofthepowerplant. By decreasing the condenser pressure from0.113barto0.029barorby 0.084 barwecanincreasepowerplantefficiency from 28.57% to 30.94%or by2.37%. Withanincreaseinpower plant efficiency of2.37%,thereispotential savingsfroma decreaseintheamountof coalusageof5.2tons/houror45,552tons /year. So that itcan increasepower plant profitsandbe moreenvironmentally friendly.

References

[1] Syed Haider Ali, Aklilu TB and Suhaimi H. Effect of Low Pressure End Conditions on Steam Power Plant Performance, (2014) MATEC Web of Conferences.

[2]Krishna,MuvvaMurali.AnalysisThe Effectof CondenserPressure (Vacuum)On Efficiency and Heat Rate Of Steam Turbine. 2016. International Journal of ScienceEngineeringandAdvance.

[3]Vasa,Ramu.PerformanceAnalysis of SteamTurbineinThermal Power Plant. 2018. International JournalofTrendy Researchin Engineering andTechnology (IJTRET).

[4] S. SenthurPrabu, M. Manichandra, G. Bhanu Prakash Reddy, D. Raghu Vamshi. "Experimental Study On Performance of Steam Condenser in 600mw Singareni Thermal Power Plant". International Journal of Mechanical Engineering and Technology (IJMET) Volume 9, Issue 3, March 2018, pp. 1095– 1106.

[5] Alus, Muammer. The Effect of the Condenser Inlet Cooling Water Temperature on the Combined Cycle Power Plant Performance. Worldwide Journal of Multidisciplinary Research and Development. 2017.

[6] Xiu Tan, Miaomiao Han. Energy Saving and Consumption Reducing Evaluation of Thermal Power Plant, IOP Conf. Series: Earth and Environmental Science 128. 2018.

[7]Ashok,J.StudyandAnalysisOnthe Performance of Condenser and CoolingTowertoForebayWater Energy. International Journalof EngineeringResearch and Modern Education (IJERME). 2016.

[8]Ion,Dosa.Efficiency Assessmentof Condensing SteamTurbine.Advances in Environment, Ecosystemsand Sustainable Tourism. 2013.

[9] Sanathara,MilanV. Parametric Analysis of SurfaceCondenser for 120 MW Thermal Power

Plant.International Journal of Engineering Research & Technology(IJERT). 2013.

[10] Ozdamar, Gokhan. Optimization of A Condenser in A Thermal Power Plant. International Conference on Heat Transfer, Fluid Mechanics and Thermodynamics. 2012.

[11] Vosough, Amir. Improvement Power Plant Efficiency with Condenser Pressure. International Journal of Multidisciplinary Sciences and Engineering. 2011.

IOP Conf. Series: Materials Science and Engineering **990** (2020) 012028 doi:10.1088/1757-899X/990/1/012028

Numerical Study for Comparison of Pseudo Modal and Direct Method in Predicting Critical Speed of Coaxial Dual Rotor System

A Sembiring[1], **A Lubis**[2]
[1]Graduate Student, Department of Mechanical Engineering, University of Lampung
[2]Department of Mechanical Engineering, University of Lampung

Email: asnawi.lubis@eng.unila.ac.id

Abstract - Machines with high speed rotation, high power, and lighter weight are the challenge in design of modern rotation machinery. In analysis of coaxial dual rotor system, the problem leads to a multi-degree of freedom system (1multi-DOF). The resulting equation of motion cannot be solved analytically. For multi-DOF system, numerical method has to be implemented. In this study, the equations of motion of co-axial dual rotor system was solved numerically using MATLAB programming language. This study compares the result of two methods used: the pseudo-modal method and the direct method, with the aim of getting the most effective method in predicting critical rotational speed by plotting Campbell diagram. It was found that the difference in the value of critical speeds between Pseudo Modal and the Direct Method is 0%-9,56% for the first 6 lowest frequencies and 0.09%-3.58% for the first 8 lowest frequencies.

1. Introduction

Machines with high speed rotational, high power, and lighter weight are the tendencies in design of rotational machinery in modern engineering. This develops in line with the increasing ability in prediction of rotor dynamic characteristics. The use of dual rotor systems (coaxial rotors) can be found in many applications such as aviation (helicopter), turbine and other fields. To get a machine with high speed, large power and light weight, the designers of dynamic rotor systems require certain techniques in predicting the dynamic characteristics of the rotor. All the dynamic characteristics of the rotor are closely related to the rotation of the rotor, so the direction of the mode is very important in the dynamic rotor. When the rotor system starts to spin, two different modes will appear, known as the forward and backward modes. The modal characteristics of a dynamic rotor system that are related to the forward and backward modes, differ from the conventional characteristics of the non-rotational structure [1]. The application of the classical modal testing method has developed well to identify the modal parameters of various structures, except for rotation machines. In analysing the dynamic characteristics of rotor, there are several methods that might be used such as the finite element method, the transfer matrix method, the time series numerical integration method and the Hubolt method for modelling the rotor shaft system and obtain faster computational time. Budiwantoro [2] and also Huang [3] investigated the dynamic characteristics of coaxial dual rotor systems using finite element method. The

[2] asnawi.lubis@eng.unila.ac.id.

Content from this work may be used under the terms of the Creative Commons Attribution 3.0 licence. Any further distribution of this work must maintain attribution to the author(s) and the title of the work, journal citation and DOI.
Published under licence by IOP Publishing Ltd

results obtained was in form of critical sonar velocity, transient displacement, and trajectory diagrams. Yanto and Hidayat [4] carried out research on the dynamic behaviour of 3D rotor shaft system (1 level) using finite element analysis and obtained results of a critical rotation variation of 192 frequencies. In an analysis of rotor dynamics, Hutahaen [5] implemented the substructure method to analyse the dynamic behaviour of the rotor (turbine 1 level). His study focused on analysing the dynamic characteristics of the rotor in form of natural frequencies plotted in Campbell diagrams. Determination of the direction of whirling orbit of turbines having high speed rotation always has its own challenges. Simulations using finite element method with the help of ANSYS software have been carried out by Jahromi [6] to determine the direction of whirling orbit and calculate the gyroscopic effect. The numerical results obtained was then verified experimentally to confirm the phenomenon of whirling orbit direction.

The pseudo modal method in a computational program was used by Zulhendri et al., to analyse the dynamic characteristics of a coaxial dual rotor system. The method was also used by Mahartana [8] for modeling and simulation of rotor balancing to see the increase in work cycle of the dynamic response.

This paper presents results of numerical study on the dynamic characteristics of a coaxial dual rotor system in the form of natural frequency (the Campbell diagram) and response to mass unbalance. This study compared the results obtained using the pseudo modal and the direct method. The computer program that has been compiled using MATLAB would then be validated using experimental studies published elsewhere.

2. Methods

By using the differential equation of motion, the dynamic characteristics of a coaxial dual rotor system can be solved. The discretization method can be used for the equation of the coaxial dual rotor system. By applying the Lagrange equation, a linear differential equation which is useful can be obtained.

2.1. Equation of Motion

The equation of motion of the rotor system can be written as follows:

$$M\ddot{\delta} + C\dot{\delta} + K\delta = F(t) \tag{1}$$

By completing Equation (1) a natural frequency can be obtained as a function of rotational speed of the rotor, Campbell's diagram, critical rotation, and excitation response, such as unbalanced mass response. For tiered rotor shaft system as shown in Figure 1, the equation of motion can be written as:

$$M\ddot{\delta} + C(\Omega_1, \Omega_2)\dot{\delta} + K\delta = F(t) \tag{2}$$

The three matrices M, C, and K are obtained by arranging matrices $M_1, M_2, C_1(\Omega_1), C_2(\Omega_2), C$ and K on bearings and seals

In a coaxial dual rotor, analysis generally leads to a multi-degree of freedom system (multi-DOF), and to solve the equation of motion, a numerical method must be used. There are two methods here used for solving the rotor system equation, namely the pseudo-modal method and the direct method.

2.2. Pseudo Modal Method

The pseudo-modal method can be used for real cases in industry. In the pseudo-modal method, the damping value C in Equation (1) is assumed to be small, so the mechanism is unknown and for that reason the damping is not installed in the rotor. Equation (1) can be defined as a modal base equation [9] as given below:

$$M\ddot{\delta} + K^{*}\delta = 0 \tag{3}$$

IOP Conf. Series: Materials Science and Engineering **990** (2020) 012028 doi:10.1088/1757-899X/990/1/012028

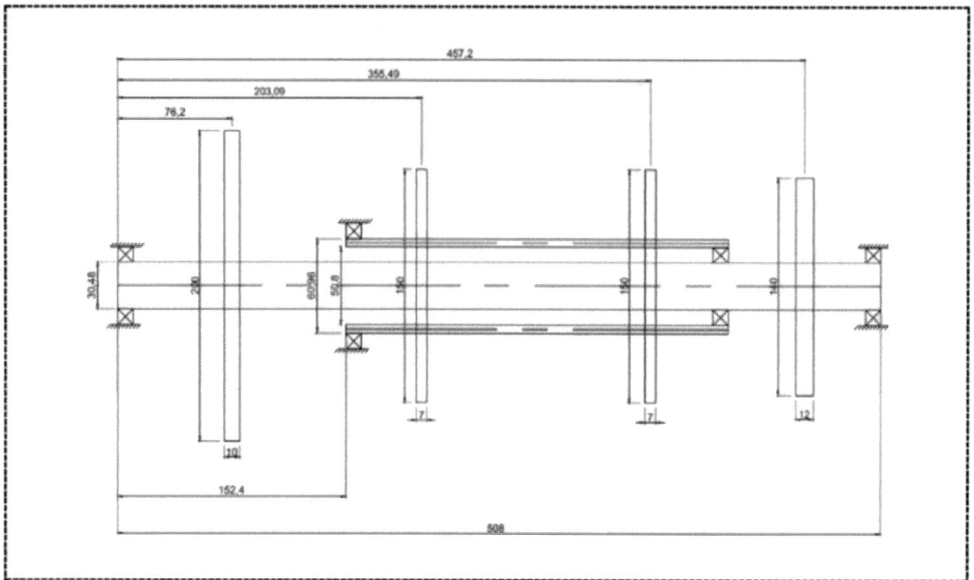

Figure 1. Coaxial dual rotor design system

The value of M is the mass matrix and K^* is the stiffness matrix obtained from the value of K, where the values of k_{zz} and k_{zx} can be ignored as indicated by the bearing being removed. The lowest number of first vibration modes (N), (N), $\phi_1, \ldots. \phi_n$ from Equation (3) can be obtained by iteration technique, and in matrix form can be written as in Equation (4):

$$\phi = [\phi_1, \ldots, \phi_n] \tag{4}$$

From Equation (3) the following new equation can be obtained.

$$\delta = \phi p \tag{5}$$

Substitute Equation (5) into the equation of motion of the rotor shaft system and multiply it by ϕ^T.

$$\phi^T M\ddot{p} + \phi^T C(\Omega)\phi\dot{p} + \phi^T K\phi p = \phi^T F(t) \tag{6}$$

The diagonal matrix $\phi^T C\phi$ is added to the modal damping c_i, which is obtained from the analogy of the single degree of freedom (SDOF) damped spring mass system:

$$c_i = 2\alpha_i\sqrt{\phi^T K\phi . \phi^T M} \tag{7}$$

with the value of the modal damping factor α_i obtained from the designer's experience. So that the solution of the equation of motion (1) without external force can be written as:

$$\phi^T M\ddot{p} + \phi^T C(\Omega)\phi\dot{p} + \phi^T K\phi p = 0 \tag{8}$$

The value of p can be assumed in the form:

$$[r^2 m + rc + k]P = 0 \tag{9}$$

The values of m, c and k can be written as:

$$m = \phi^T M\phi = diag\{\phi_i^T M\phi\} \tag{10}$$

$$c = \phi^T C\phi = diag\{\phi^T C\phi\} + C_i \tag{11}$$

$$k = \phi^T K\phi = diag\{\phi_i^T K^*\phi_i\} + \phi_i^T K^{**}\phi_i \tag{12}$$

The stiffness K^{**} is expressed as:

$$K^{**} = K - K^* \tag{13}$$

Equation (9) can be written in the form of a matrix equation, as in Equation (14).

$$\begin{bmatrix} 0 & I \\ -k^{-1}m & -k^{-1}c \end{bmatrix} \begin{Bmatrix} rP \\ P \end{Bmatrix} = \frac{1}{r} \begin{Bmatrix} rP \\ P \end{Bmatrix} \tag{14}$$

The solution to the problem of eigenvalue or natural frequency and vibration mode of equation (14) can be obtained in complex quantities.

$$r = \frac{-\alpha_i \omega_i}{\sqrt{1-\alpha_i^2}} \tag{15}$$

The value of ω_i is a natural frequency and α_i is a viscous damping factor. If equation (15) > 0, the system is not stable. From the results obtained, Campbell's diagram can be drawn showing critical areas so that the critical velocity can be known. Unbalanced mass response may be obtained from the following equation:

$$m\ddot{p} + c\dot{p} + k\delta = f_2 \sin \Omega t + f_3 \cos \Omega t \tag{16}$$

with the values f_2 and f_3 can be expressed as follows:

$$f_2 = \phi^T F_2 \tag{17}$$

$$f_3 = \phi^T F_3 \tag{18}$$

Assuming an answer for under-damped vibration as follows:

$$p = p_2 \sin \Omega t + p_3 \cos \Omega t \tag{19}$$

By substituting equation (19) into equation (16) the following matrix form will be obtained:

$$\begin{bmatrix} k - m\Omega^2 & -\Omega c \\ \Omega c & k - m\Omega^2 \end{bmatrix} \begin{Bmatrix} p_2 \\ p_3 \end{Bmatrix} = \begin{Bmatrix} f_2 \\ f_3 \end{Bmatrix} \tag{20}$$

The solution of equation (20) and from equation (19), the following displacement vectors are obtained:

$$\delta = \phi\{p_2(\Omega) \sin \Omega t + p_3(\Omega) \sin \Omega t\} \tag{21}$$

2.3. Direct Method

The difference between the pseudo modal method and the direct method in solving the coupled equation of motion is that there is no equation reduction in the direct method. In the direct method, the answers are considered in the form of equation below:

$$\delta = \Delta e^{rt} \tag{22}$$

When Equation (22) and its derivative are then substituted into Equation (2), the following equation is obtained.

$$(r^2 M + rC + K)\Delta = 0 \tag{23}$$

Equation (23) can be written in the form of state space.

$$\begin{bmatrix} 0 & M \\ M & C \end{bmatrix} \begin{Bmatrix} r\Delta \\ r \end{Bmatrix} = \frac{1}{r} \begin{bmatrix} M & 0 \\ 0 & -K \end{bmatrix} \begin{Bmatrix} r\Delta \\ r \end{Bmatrix} \tag{24}$$

Equation (24) is eigenvalue problem. The first two eigenvalues and corresponding eigenvector can be obtained in complex quantities as given in Equation (25) and (26) respectively:

$$r_1 = \frac{\alpha_1 \omega_1}{\sqrt{1-\alpha^2}} \pm j\omega_i \tag{25}$$

The eigenvector values can be seen in the following equation.

$$\Delta_1 = R_1 \pm jL_1 \tag{26}$$

Campbell's diagram of the natural frequency of the system can be obtained by solving those equation.

3. Results and Discussion

Equation of motion of coaxial dual rotor system can be expressed as in Equation (2). The matrix values $\boldsymbol{M}$, $\boldsymbol{C}$, and $\boldsymbol{K}$ in Equation (2) are obtained by arranging the matrix $\boldsymbol{M_1}$ (disk mass matrix and inner axis), and $\boldsymbol{M_2}$ (disk mass matrix and outer shaft). While the value of matrix $\boldsymbol{C}$ is arranged based on the gyroscopic effect of the disk, the shaft (inside and outside), and the effect of bearings and seals attenuation. The $\boldsymbol{K}$ matrix is obtained from the axial force matrix (inside and outside), as well as bearings and seals thickness. The form of Equation (2) can be arranged into the global equation matrix as follows:

$$\begin{bmatrix} M_1 & 0 \\ 0 & M_2 \end{bmatrix}\ddot{\delta} + \begin{bmatrix} C_1\Omega_1 & 0 \\ 0 & C_2\Omega_2 \end{bmatrix}\dot{\delta} + \begin{bmatrix} K_1 & 0 \\ 0 & K_2 \end{bmatrix}\delta = F(t) \tag{27}$$

The coaxial dual rotor is a system that has two shafts that coincide on both axes and experience rotation simultaneously with different or equal rotational speeds. The two shaft systems are connected by intershaft bearings as shown in Figure 1. The model of the graded rotor shaft system can be seen in Figure1. Data and dimensions of the coaxial dual rotor system are given in Tables 1 and 2, characteristics of bearings and seals are given in Table 3, and material properties of the shaft are given in Table 4:

Table 1. Dimension of the shaft

Shaft	Inner Diameter (mm)	Outer Diameter (mm)	Length (mm)
Inner shaft (1)	0	30,48	508
Outer shaft (2)	50,8	60,96	254

Table 2. Rotor data

Rotor	D_1	D_2	D_3	D_4
$M(kg)$	10,51	7,01	3,5	7,01
$I_{Dx}10^{-2}(kgm^2)$	4,295	2,14	1,35	3,39
$I_{Dy}10^{-2}(kgm^2)$	8,59	4,29	2,71	6,78

Table 3. Characteristics of bearings and seals

Bearings and seals	1	2	3	4
$k_{xx} = k_{zz}, (N/m)$	$2{,}63 \times 10^7$	$1{,}75 \times 10^7$	$0{,}875 \times 10^7$	$1{,}75 \times 10^7$

Table 4. Material properties of the shaft

Density (ρ)	Elastic modulus (E)	Poison ratio (v)
7800 kg/m^3	2×10^{11} N/m^2	0,3

By using the finite element method to model the system, the rotor shaft system is divided into 7 elements (inside) and 4 elements (outside), so that there are 11 elements in modelling coaxial dual rotor system as shown in Figure 2.

IOP Conf. Series: Materials Science and Engineering **990** (2020) 012028 doi:10.1088/1757-899X/990/1/012028

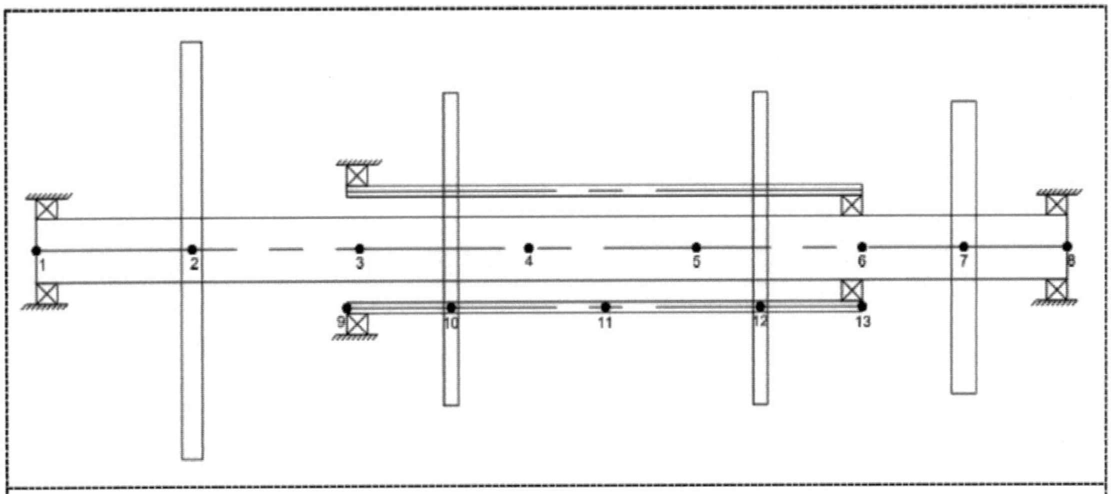

Figure 2. Finite Element Modelling of coaxial dual rotor system

The location of nodes on the inner and outer shaft are given in Tables 5 and 6 respectively. Data for finite element is given in Table 7.

Table 5. Nodal coordinates on the inner shaft

Nodal	1	2	3	4	5	6	7	8
Absis (mm)	0	76,2	158,7	241,3	323,8	406,4	457,2	508

Table 6. Nodal coordinates on the outer shaft

Nodal	9	10	11	12	13
Absis (cm)	152,4	203,2	279,4	355,6	406,4

Table 7. Finite element modeling node coaxial dual rotor system

No	Elemen		Node
1	Shaft	Inner	1-8
		Outer	9-13
2	Rotor	Inner	2 and 7
		Outer	10 and 12
3	Bearings and seals	Inner	1 and 8
		Outer	9
		Intershaft Bearings	13

Numerical studies were carried out to compile a program used in predicting the dynamic characteristics of coaxial dual rotor system. In this research, the program was written using MATLAB software. Two solution methods were written for solving the equation of motion: the the pseudo modal and the direct method

IOP Conf. Series: Materials Science and Engineering **990** (2020) 012028 doi:10.1088/1757-899X/990/1/012028

The speed rotation was in the range of 0 to 14,000 rpm. The first six lowest frequencies computed using the pseudo modal and the direct method is shown in Figure 3 and 4 respectively. Figure 5 and 6 show the first eight lowest frequencies obtained using the pseudo modal and the direct method respectively.

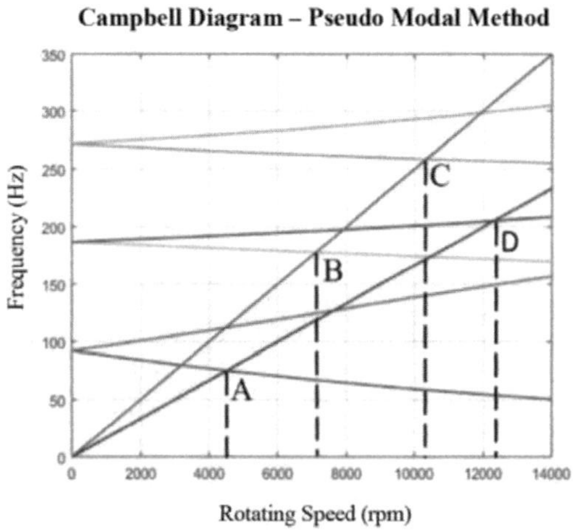

Figure 3. Campbell Diagram – Pseudo Modal Method with six lowest frequencies.

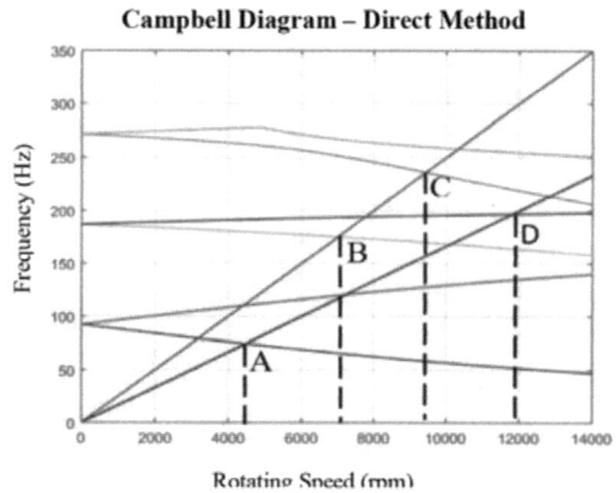

Figure 4. Campbell Diagram - Direct Method with six lowest frequencies.

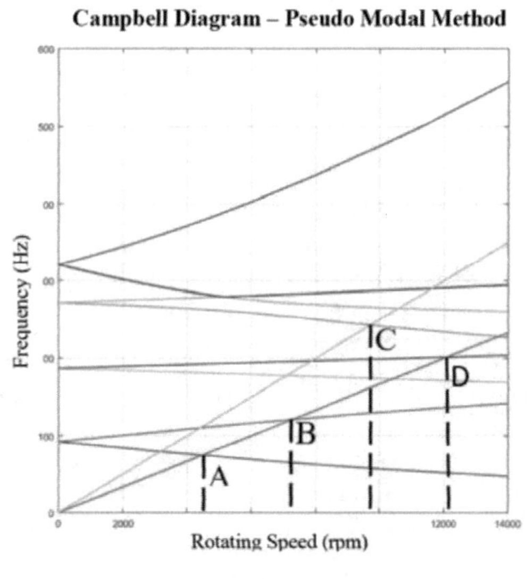

Figure 5 Campbell Diagram - Pseudo Modal method with 8 lowest frequencies

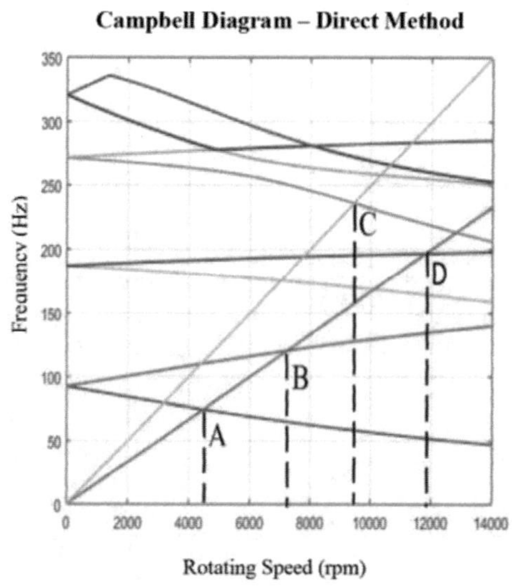

Figure 6 Campbell Diagram - Direct method with 8 lowest frequencies.

The values of Critical speed obtained using the Pseudo Modal method and the Direct Method are given in Table 8 for the first six lowest frequency and in Table 9 for the first eight frequency.

Table 8. Various critical speed values (6 lowest frequency).

Coordinate	Pseudo Modal Method (rpm)	Direct Method (rpm)
A	4.475	4.475
B	7.118	7.014
C	10.341	9.439
D	12.307	11.810

Table 9. Various critical speed values (8 lowest frequency).

Coordinate	Pseudo Modal Method (rpm)	Direct Method (rpm)
A	4.479	4.475
B	7.265	7.014
C	9.726	9.439
D	12.082	11.810

It can be observed from Table 8 and 9 that there are difference in values of critical rotation speed between the first six lowest frequency and the first eight lowest frequency if the Pseudo Modal method is used. This is due to using the Pseudo Modal method, the value of the matrix damping factor C is not included in the global equation system as shown in Equation (2). This causes the critical rotation speed value obtained by the Pseudo Modal method to be higher when compared to the critical rotation speed value obtained using the Direct method.
The percentage change in the value of critical speed using the Pseudo Modal method between the first six lowest frequencies and the first eight lowest frequencies are given in Table 10.

Table 10. Changes in the critical speed value of the Pseudo Modal method

Coordinate	Pseudo Modal Method (rpm)	
	6 frequency	8 frequency
A	4.475	4.479
B	7.118	7.265
C	10.341	9.726
D	12.307	12.082

If the results obtained using the Pseudo Modal method are compared with the results obtained using the Direct method, the results can be seen in Table 11 for the first six lowest frequencies, and Table 12 for the first eight lowest frequencies.

Table 11. The difference in values of critical rotation speed between the Direct method and the Pseudo Modal method for the first six lowest frequencies.

Coordinate	Pseudo Modal Method (rpm)	Direct Method (rpm)	Percentage Difference
A	4.475	4.475	0%
B	7.118	7.014	1,48%

C	10.341	9.439	9,56%
D	12.307	11.810	4,21%

Table 12. The difference in values of critical rotation speed between the Direct method and the Pseudo Modal method for the first eight lowest frequencies

Coordinate	Pseudo Modal Method (rpm)	Direct Method (rpm)	Percentage Difference
A	4.479	4.475	0,09%
B	7.265	7.014	3,58%
C	9.726	9.439	3,04%
D	12.082	11.810	2,30%

It should be noted that the Pseudo Modal method in the process of solving the equations of the rotor shaft system uses the number of modes (NoM) vibrations that want to be analyzed. Meanwhile, the Direct method uses the maximum number of elements (NMAX), on the rotor shaft. These differences can be seen in the following MATLAB program script.

```
for k=1:NoM
    PSI(:,k)=V(:,No_urut(k));
end
mr=PSI.'*M*PSI;
kr=PSI.'*K*PSI;
f2=PSI.'*F2;
f3=PSI.'*F3;
f=[f2
    f3];
ci=2*Alfai*sqrt(kr*mr);
cidum=zeros(NoM);
```

Figure 7. Script for solving global matrix equations (Pseudo Modal Method)

```
for k=1:NMAX
        k1=2*k-1;
        No_urut(k)=dum1(k1,2);
    end
    for k=1:NoM
Fn_dum1(k)=Fn_dum(No_urut(k));
```

Figure 8. Script for solving global matrix equations (Direct Method)

4. Conclusion

The Pseudo Modal method and the Direct Method can be used to predict the critical rotation speed of a coaxial dual rotor shaft system. By using the Pseudo Modal method, the value of the damping matrix C is considered small so that it was not taken into account in the completion of the motion system. Assuming small value of C matrix, the value of critical speed obtained using the Pseudo Modal method is greater than those obtained using the Direct method. There is a difference in the value of critical speed using the Pseudo Modal method between the number of vibrational mode (NoM) between the first six lowest frequencies and the first eight lowest frequencies. Further research is needed on experimental study of coaxial dual rotor shafts to validate the best results between the Pseudo Modal method and the Direct Method.

Acknowledgment

The authors are very grateful to the Research Institute (LPPM) of the University of Lampung for funding this research under the scheme of Graduate (Pascasarjana) Research, FY 2020.

IOP Conf. Series: Materials Science and Engineering **990** (2020) 012028 doi:10.1088/1757-899X/990/1/012028

5. Reference

[1] Jei, Y. K 1993 Journal of Vibration and Acoustics **115** 165-176.
[2] Budiwantoro, B 2001 Jurnal Teknik Mesin **XVI** 32-42
[3] Huang, Y. C.-M 2001 Journal of Vibration and Acoustic **123** 428-434.
[4] Yanto, A., dan Hidayat, R 2014 Jurnal Teknik Mesin **4** 75-83.
[5] Hutahean, R 2014 Jurnal Ilmiah Teknik Mesin Cylinder **1** 39-56.
[6] Jahromi, A. F.-F 2015 Springer International Publishing. Switzerland 879-887..
[7] Zulhendri, H., Sembiring, A., Lubis, A 2018 Prosiding SNTTM XVII Kupang. Universitas Nusa Cendana 029-038.
[8] Mahartana, IB 2017 Jurnal Teknik ITS **6** F120-F125.
[9] Lalanne, M., and Ferraris, G 1990 *Rotordynamics Prediction in Engineering*. John Wiley and Sons Inc., New York.

IOP Conf. Series: Materials Science and Engineering **990** (2020) 012029 doi:10.1088/1757-899X/990/1/012029

Kinematic analysis on four-bar mechanism model using PID Controller

R Arman[1]
[1]Mechanical Engineering Department, UniversitasBung Hatta
Jl. Gajah Mada No. 19 OloNanggalo, Padang, 25143, Indonesia

E-mail: rizky.arman@bunghatta.ac.id

Abstract. A four-bar mechanism is used in most mechanical equipment to achieve the process or movement. Such mechanisms consist of linksthat move relative to one another. To identify kinematics analysis of a mechanism such as displacement and angular velocity of a four-bar mechanism in one positiongraphically.The analysis conducted throughout the four-bar rotating in 360^0will take a lot of time. For such cases, a computer-assisted analytical is the right way of settlement. The analyses were performed using the Matrix Laboratory software, and establish a four-bar mechanism as a case of study research.Controller PID (Proportional Integral Derivative) is one type of regulator that is widely used, the characteristics of each P,I, and D controllers as well as how to use it to get the desired response. The analysis was made on the variation of rotation and crank length, while the dimension of the link and follower circuit is considered permanent. Comparison of angular velocity for varying stem lengths can conclude that if there is a change in input bar length, it will affect the angular velocity. The greater the length of the inputlink and so thattheangular velocity would begettinglower.

Introduction

A kinematic chain is a system of linking rods in the form of rigid objects that are joined together or in a state of mutual contact so that it is possible for them to move relative to one another. If one of the links is fixed and the movement from the other connecting rod to its new position will cause each other link to move to certain positions the system is predicted to be a restricted kinematic chain.If one of the linking rods is kept fixed from the movement of the other connecting rod to its new position, it will not cause each other connecting rod to move to a certain position that has been predicted thus the system is an infinite kinematic chain.

Analysis of the displacement and angular velocity of the four-bar mechanism, usually done graphically. To identify the displacement and angular velocity of the four-bar mechanism in one position graphically can be done simply and quickly. However, if the analysis is carried out to identify the displacement and velocity of the four-bar mechanism throughout the rotating position of the360^0 input bar, graphically it will take a lot of time. For such cases, the method of analytic with the help of computers is the right solution.The four-bar mechanism is used in some mechanical equipment to achieve certain processes or movements. Such a mechanism consists of rods that move relatively one against the other. In general notation, the bar is also referred to as a link. The two links that allow

[1] Rizky.arman@bunghatta.ac.id.

Content from this work may be used under the terms of the Creative Commons Attribution 3.0 licence. Any further distribution of this work must maintain attribution to the author(s) and the title of the work, journal citation and DOI.
Published under licence by IOP Publishing Ltd

IOP Conf. Series: Materials Science and Engineering **990** (2020) 012029 doi:10.1088/1757-899X/990/1/012029

relative movement between are called joints. Thus the mechanism of the four-bar in question consists of four links that are connected by joints in such a way as to allow for relative movement between the existing links[1,2,3].

Material And Methods

In studying the movements of the machine elements, we usually draw these parts in the form of sketches so that only the parts that will give effect to the movement are noticed.PID Controller (Propotional Integral Derivatives) is one type of regulator that is widely used. In addition, this system is easily combined with other setting methods such as *Fuzzy* and *Robust*. To make a better regulatory system, this paper is limited to systems with the *Unity Feedback System*, as shown in Figure 1[4,5].

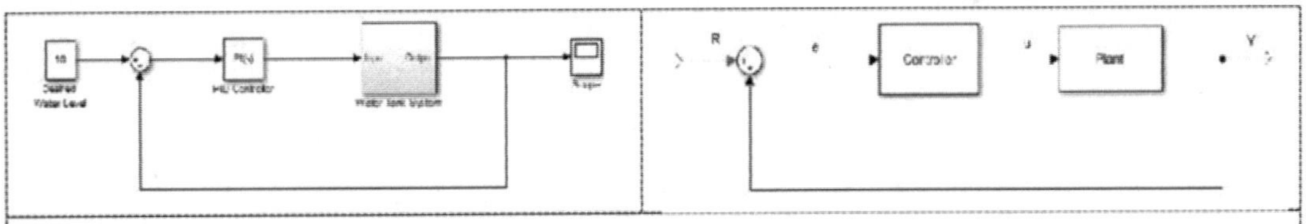

Figure 1. Block diagrams of *Unity Feedback System.*

First, consider the workings of the PID controller in the closed-loop system using the scheme shown above. Variable (*e*) describes tracking error, different input value (*R*), actual output (*Y*). This signal error is sent to the PID controller, and the controller will calculate the overall derivative of the integral signal error. The signal (*u*) that has passed through the controller, now equals the proportional gain (*Kp*) multiplied by the size of the error plus the reinforcement integer (*Ki*) multiplied by the size of the integral error plus the derivative reinforcement (*Kd*) multiplied by the derivation error[6,7].

$$\mu = K_p e + K_i \int e \, dt + K_D \frac{de}{dt} \quad (1)$$

Signal (*u*) will be sent to the plant and will get a new output (*y*). This new output (*y*) will be sent back to the sensor to look for new signal errors (*e*). The controller brings the new signal error and calculates the derivatives and the integrals once again. The process will run continuously as before. The proportional controller (*Kp*) will have the effect of reducing incremental time but does not erase steady-state errors. The integral controller (*Ki*) will have the effect of erasing steady-state errors, but resulting in deteriorating transient responses[8]. The derivative controller will affect increasing system stability, reducing overshoot and increasing transfer response. The effects of each controller (*Kp, Kd, Ki*) in the closed-loop system are shown in Table 1[9].

Table 1. Controller closed loop system

Closed Loop Response	Incremental Time	Overshoot	Decremental Time	Steady State Error
Kp	decrease	increase	small shifting	Decrease
Ki	decrease	increase	increase	Shrink
Kd	small shifting	decrease	decrease	small shifting

Note that the correlation relationship may not be entirely accurate, because *Kp*, *Ki* and *Kd* are mutually independent. Changing one variable an change the other two. For this reason, Table 1 above is only used as a reference when we determine the values for *Ki*, *Kp*, and *Kd*. From table 2, it is known that the proportional controller (*Kp*) reduces incremental time, increases overshoot and reduces steady-state errors[10]. The closed-loop transfer function of the system above with proportional control is;

IOP Conf. Series: Materials Science and Engineering **990** (2020) 012029 doi:10.1088/1757-899X/990/1/012029

$$\frac{X(s)}{F(s)} = \frac{K_p}{s^2+10s+(20+K_p)} \tag{2}$$

To get information regarding the solution to this problem, a literature study was conducted. Information regarding this problem was obtained from books and journals relating to the four-bar mechanism in general obtained from various sources. At this stage, a data analysis mechanism for the four-bar is made[11,12,13]. The method of this research is the study of kinematics on the four-bar mechanism. Supported by establishing data used to execute numerical commands in Matlab of the four-bar mechanism includes:

a. *Input data in the Windows command*. Matlab Command windows/edit is a window that is opened when Matlab is run, in the windows command the commands entered are typed and can be saved using the command record.

b. *Simulink mechanism for four-bar*. The mechanism of the four bars in Matlab can be seen using Simulink, in the corner Simulink of the four-stem mechanism that we analyze, the data released is in the form of moving images.

c. *Data analysis*. Data analysis in Matlab can be seen by outputting the graph with the help of windows figures. Output data can be in the form of a graph.

d. *The output from Matlab Analysis*. The output from Matlab can be Simulink and graph.

The stages of the simulation model begin with making a PID diagram in the Matlab software. Overall, the display of the PID diagram in the four-bar mechanism Simulink can be seen in Figure 2. In Figure 2, we see a four-bar mechanism, in which we have to display the Simulink menu located at the top of the menu when the Matlab software opened. After Simulink appears, we select *File >> New >> Blank Model*.

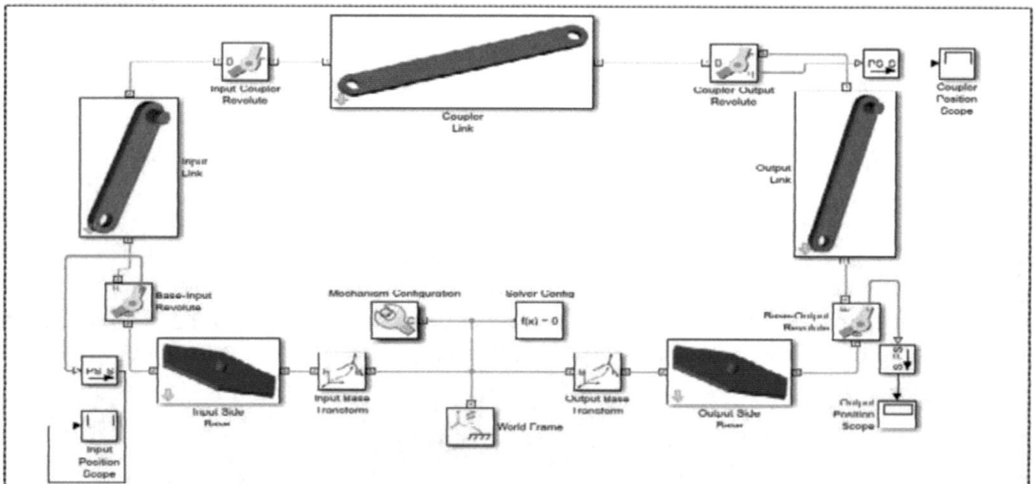

Figure 2.PID block diagrams of four-bar mechanism

Result And Discussions

In this study, a simulation was performed for variations in the speed and crank length and then analyzed the kinematic response of the four-bar mechanism. Simulation is a technique that imitates operations or processes that occur in a system with the help of computer devices and is based on certain assumptions so that the system is studied scientifically. By making a model of a system, it is expected to be easier to carry out an analysis. This is a modeling principle, namely that modeling aims to facilitate analysis and development. Modeling is one way to study the system and the model itself and also the differences in behavior, as shown in Figure 3.Subsequent, we will compile a four-bar PID diagram. First, we have to look for the Simulink Library, in the Simulink library there are various kinds of items to design the PID controller[14].

To enter the PID block diagram into the Simulink blank model, it is done by dragging the block with the crucible and directed to the blank Simulink model.In this stage, we will analyze the model in each of the links above by changing the size of the crank length and the variation of rotation. The modified data is contained in the block parameters in each link. The output is in the form of a graph, so we can compare the output graph with different input data. For standard sizes on each link must be set, if the link size does not match the standard, then Simulink cannot operate normally. The standard size for the link can be seen in Table 2.

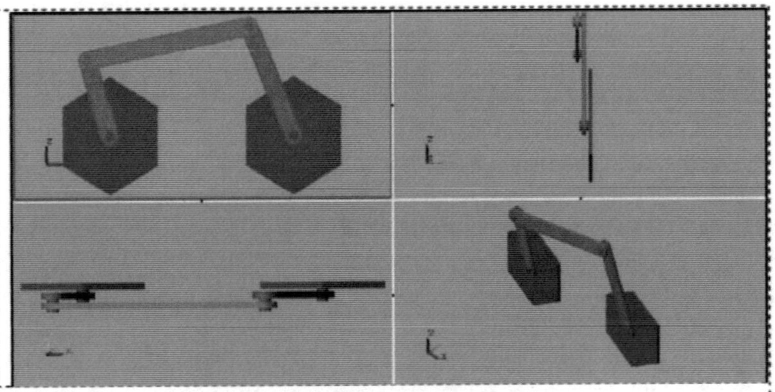

Figure 3. Four-bar mechanism model.

Table 2. Standart properties and sizes of each link with variation

Block Parameter	Input Link	Coupler Link	Output Link
Link Density	10 ~ 12 kg/vol	10 ~ 12 kg/vol	12 ~ 14 kg/vol
Link Length	8 ~ 12 cm	18 ~ 22 cm	12 ~ 16 cm
Link Width	2 ~ 6 cm	2 ~ 6 cm	2 ~ 6 cm
Link Thickness	0.5 cm	0.5 cm	0.5 cm
Hole Radius	0.5 cm	0.5 cm	0.5 cm

If standard sizes and variations have been set, Simulink can move normally. Following that, we can also see the output window in the analysis table model of the four-bar mechanism from the input link, coupler link and output link, as seen in the graph of Figure 4.

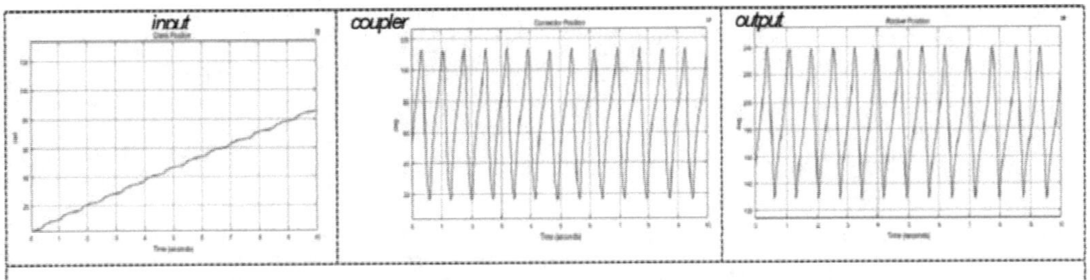

Figure 4. Graph of input, coupler and output link position

When we know the angular velocity, we can determine the linear velocity of the input rod. And the linear velocity for this input link is constant. The following in Figure 5 is a trajectory velocity simulation model for the four-bar mechanism with the number of cycles 14, 16 and 19 in the input link[1,2,8].

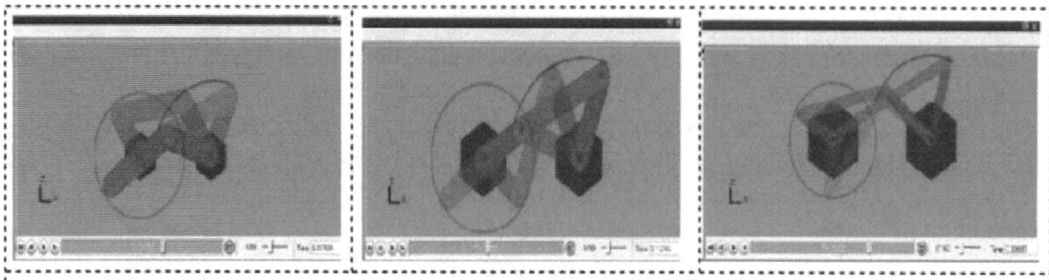

Figure 5. Trajectory velocity of simulation model in the four-bar mechanism at 14, 16 and 19 cycle

Conclusion

From the mechanism that has been simulated by the PID controller method, it is obtained that the angular velocity value of the mechanism for the link length varied is; for the input rod length of 8 cm, the angular velocity is 119.38 rad/s and the linear velocity is constant at 95.44 cm/s. Whereas for the input rod length of 10 cm, the angular velocity is 100.48 rad/s and the linear velocity is constant at 100.4 cm/s. Meanwhile, for the input rod length of 12 cm, the angular velocity is 87.92 rad/s and the linear velocity is constant at 105.48 cm/s. So from the ratio of angular velocity to varying rod length, it can be concluded that if there is a change in the length of the input link, it will affect the angular velocity. The larger the input link length, the lower the angular velocity.

References

[1] D. H. Myszka, *MACHINES AND MECHANISMS Fourth Edition*. .
[2] J. Gallardo-Alvarado, *Kinematic Analysis of Parallel Manipulators by Algebraic Screw Theory*. Springer International Publishing, 2016.
[3] R. Arman, A. Isra Mahyuddin, W. Marthiana, and I. Satria, "Experimental Investigation of Vibration Response of a Flexible Coupler In a Four Bar Mechanism Due to Varying Crank Length and Crank Speed," in *MATEC Web of Conferences*, 2018, vol. 248.
[4] D. Hroncová, I. Delyová, and P. Frankovský, "Kinematic Analysis of Mechanisms Using MSC Adams," no. August 2014, 2016.
[5] S. Kumar, B. Bongardt, M. Simnofske, and F. Kirchner, "Design and Kinematic Analysis of the Novel Almost Spherical Parallel Mechanism Active Ankle," pp. 303–304, 2019.
[6] K. Ayusawa, "Comprehensive theory of differential kinematics and dynamics towards extensive motion optimization framework," 2018.
[7] L. J. Ricalde, M. Flota-ban, and R. Peo, "On the coupler point velocity control of variable input speed servo-controlled four-bar mechanism," vol. 8, no. 11, pp. 1–9, 2016.
[8] M. Ning *et al.*, "Design and analysis for a multifunctional rescue robot with four-bar wheel-legged structure," vol. 10, no. 2, pp. 1–14, 2018.
[9] "Application of moving sliding mode control for a DC motor driven four-bar mechanism," vol. 10, no. 3, pp. 1–13, 2018.
[10] H. Kim, Y. Jeong, J. Kim, and T. Nam, "A prototyping tool for kinetic mechanism design and fabrication : Developing and deploying M . Sketch for science , technology , engineering , the arts , and mathematics education," vol. 10, no. 12, pp. 1–16, 2018.
[11] S. Lin, S. Wang, C. Wang, L. Zhang, and Y. Zhai, "Kinematics and dynamics analysis of a novel 2PC-CPR parallel mechanism," no. October, pp. 1–10, 2016.
[12] Q. Xu, Y. Yang, Z. Jing, and S. Hu, "Forward kinematics analysis for a class of asymmetrical parallel manipulators," no. February, pp. 1–12, 2017.
[13] S. Staicu, Z. Shao, Z. Zhang, X. Tang, and L. Wang, "Kinematic analysis of the X4 translational – rotational parallel robot," no. October, pp. 1–12, 2018.

[14] X. Liu, X. Zheng, and S. Li, "Development of a humanoid robot hand with coupling four-bar linkage," vol. 9, no. 1, pp. 1–13, 2017.

IOP Conf. Series: Materials Science and Engineering **990** (2020) 012030 doi:10.1088/1757-899X/990/1/012030

The Utilization of Hydrodinamics Models in Validating the East Java Rip Current in the Era of Industrial Revolution 4.0

S Hermawan[1]*, K Harsono[1], N Bong[1], D Gho[1]
[1]Department of Engineer Profession Program and Civil Engineering Department, Petra Christian University, Surabaya, East Java, Indonesia

E-mail: shermawan@petra.ac.id

Abstract. The earth has an ocean area larger than the land area so that there are many beaches found throughout the world, including Indonesia. Behind its beauty, there is one phenomenon that endangers tourists when doing activities on the near shore area, namely: rip current. For example, the incident at the research location, namely Klayar Beach, East Java, on January 2, 2015, claimed lives. Therefore, this study aims to validate the existence of rip current in East Java, especially at the research location by utilizing hydrodynamic modeling in the era of the industrial revolution 4.0 based on data, internet, and technology as well as human resources during this corona pandemic. There are 3 stages of the methodology in this research, namely data collection and processing, modeling, and modeling validation. The outcomes of this research prove that the phenomenon of rip current with rip current speeds occurs in the range 0.16 - 0.18 m/s which causes 2 tourists to die. It is hoped that the results of this research can be used as input for the stakeholder concerned for the development of coastal tourism in coastal areas on the south coast of East Java in making policies.

1. Introduction

Geographically, the Earth has a greater ratio of sea area than land area. With such geographical conditions, it is not surprising many beaches around the world, one of which is in Indonesia. Indonesia has a total of 17,504 islands with a sea area of 5.8 million km² with a coastline of 95,161 km, the second-longest in the world[1]. The beach is the border area between the land and the ocean and this area is categorized as a natural tourist spot because it attracts tourists to visit. Behind all the beauty of beach tourism, it turns out that there have been many casualties, especially due to the rip current phenomenon. For example, in Australia, 142 cases of tourists died from being dragged down by rip current[2]. In Indonesia this also often happens, such as the incident on January 2, 2015, 2 tourists died on Klayar Beach, East Java. Rip current itself is backflow from the coast to the sea which is known to have a high speed [3], where rip itself means to tear or pull something quickly, forcibly away from something. Rip current is dangerous because it can 'eliminate' a person's swimming ability when trapped in it[4]. There are still many Indonesian people who are not aware of the rip current The occurrence of a rip current is caused by high wind speed and duration which propagate high wave, along with of the near shore current as well as long shore current. Due to bathymetry condition, the long shore current meet at specific location which trigered current backflow to the sea, known as rip current.

[1] shermawan@petra.ac.id

Content from this work may be used under the terms of the Creative Commons Attribution 3.0 licence. Any further distribution of this work must maintain attribution to the author(s) and the title of the work, journal citation and DOI.
Published under licence by IOP Publishing Ltd

The collection of data in this research is helped by the prevalence of industrial revolution 4.0 technology, including human resources, data, and internet connection.[5]. Other tools and software, such as sonar, 3D delft software, dashboard delft software, wind rose software, and others also assisted in data retrieval and processing. The ability to collect real data, reflecting the actual situation in the field, enabled the hydrodynamic models to simulate the rip current phenomenon. This study aims to create a hydrodynamic model whose results are closest to the actual situation on the East Java coast. Researchers intend to be able to predict the occurrence of the rip current so that it can be anticipated before it takes a new toll.

2. Methodology

2.1 Research location

As can be seen in Figure 1, the research location is at Klayar Beach (longitude position: East 110°56'52.24" and latitude position: South 8°13'26.27"), Pacitan Regency, East Java. This area is a famous and favorite recreational area where located about 35 km from the capital of Regency along with having cars type of topography.

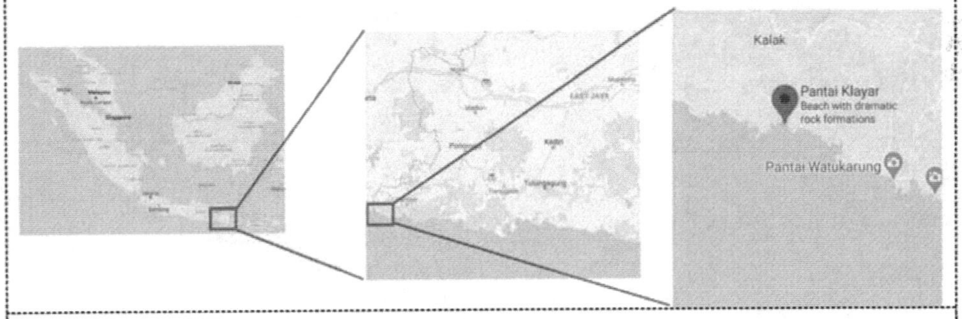

Figure 1. Research location

2.2 Data collection

To perform this modeling required wind data. Wind data is used to be able to analyze wave height taken from March 2010 to March 2020. The wind data grouping is based on the percentage of wind speed and direction recorded by the Meteorology, Climatology, and Geophysics Agency (BMKG). The wind data used were taken from the closest station to the research location, namely the Sleman Geophysical Station with the ID World Meteorological Organization (WMO) 96855 and at latitude 07 ° 49'12 ", longitude 110 ° 18'00". For bathymetry data taken from Gebco for offshore areas and near shore areas because it is not possible to collect primary data in this pandemic condition.

2.3 Hydrodinamic model determination

In this study, the hydrodynamic calculations were carried out at the research location manually along with the hydrodynamic model developments[6,7]. The first modeling is done to calculate numerical manually the wave height using empirical methods Sverdrup Munk and Breitsneider or SMB [8] and Darby Shire method using graphical method. While the calculation of wave height using the hydrodynamic model method using Delft Dashboard software is expected to show several simulation results such as water level, wave height, and current [9–12]

The making of this hydrodynamic model is trial and error to find the closest result to the actual conditions. From several trials and errors that were tested, finally found a model that was close to the situation in the field. By using a 2,200-meter square grid, manning roughness 0.05, time step 60 minutes, wind data, 2 observation points at Klayar, and Benoa Beach. Depth data comes from GEBCO at the position of Universal Transverse Mercator (UTM) 49L which will be simulated from 1-31 March 2020 as can be seen in Figure 2.

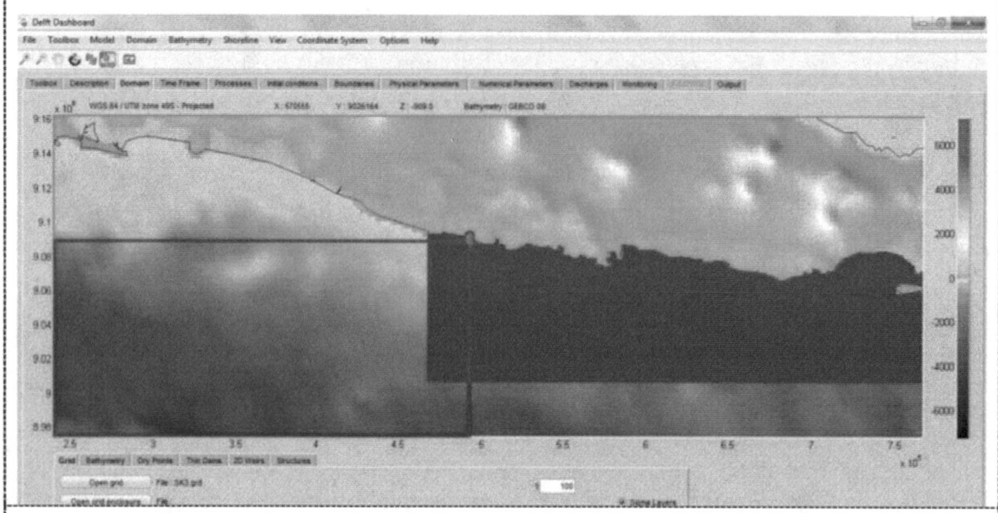

Figure 2. Hydrodynamic numerical modeling with DelftDashboard

3. Results and Discussions

3.1 High of water level

From Figure 3a, it is found that regarding hydrodynamics model outcomes, the average high water level at the research location (Mean Sea Level) at maximum tide is 1.02 meters and the maximum low tide is -0.9 meters at the Klayar observation point. Furthermore, in Figure 3b it can be seen that the simulation results must be tested with validation. Validation for MSL was carried out at the nearest Klayar observation point by finding the smallest root mean square error (RMSE) simulation results with readings from IOC-sea level monitoring in Klayar. The comparison results get an RMSE value of 0.16 meters, which means that the modeling can be said to be close to the actual situation so that the MSL at Klayar Beach can also be said to be close to the actual situation.

IOP Conf. Series: Materials Science and Engineering **990** (2020) 012030 doi:10.1088/1757-899X/990/1/012030

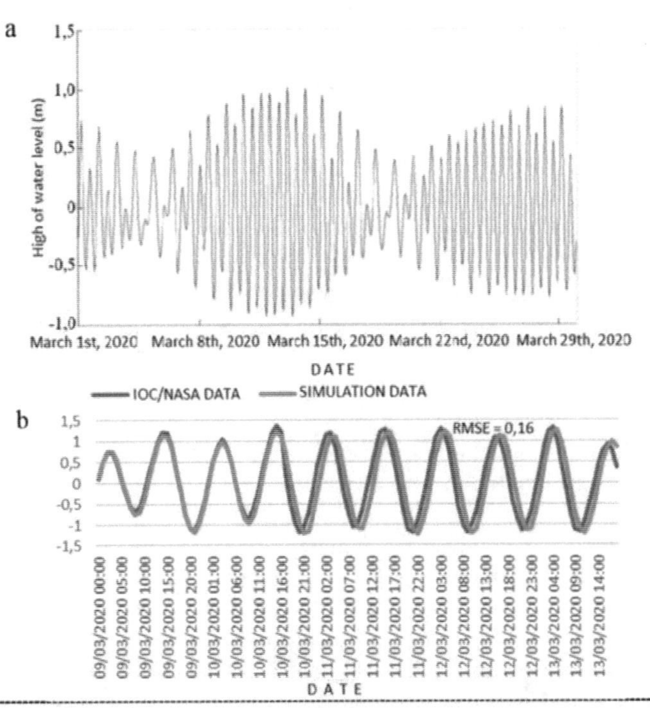

Figure 3. Results of mean sea level and validation at the Klayar Beach observation points

3.2 Wave height

The results of the second hydrodynamic modeling produce wave height which can be seen in Figure 4. The wave height at Klayar Beach shows a range of 3.4 meters. To validate wave height, it can be done in 2 ways, namely the empirical method (Sverdrup Munk and Breitsneider or SMB) and graph reading (Darby Shire). Calculation of wave height using hydrodynamic modeling, empirical methods, or graphic methods or the Darby Shire all show a range of 3.4 meters. From the graphical results of the nomogram or Darby Shire, it shows a wave height of 3.4 meters. Thus, it is proven that the hydrodynamic model simulation using Delft dashboard, graphical and empirical models provides appropriate results.

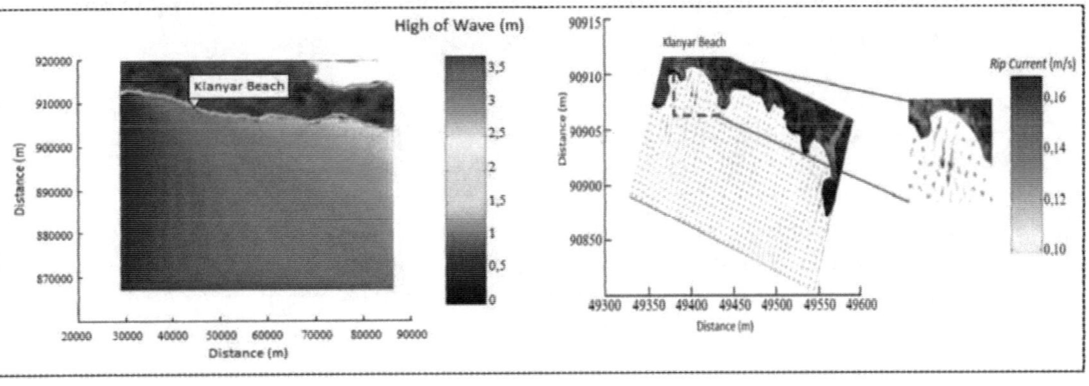

IOP Conf. Series: Materials Science and Engineering **990** (2020) 012030 doi:10.1088/1757-899X/990/1/012030

Figure 4. Simulation results of wave height at Klayar Beach (UTM 49L)

Figure 5. Simulation results of Wave height at Klayar Beach(UTM 49L)

3.3 Rip current

Then, the last outcome due to the hydrodynamic model is to read the rip current speed, it can be done by paying attention to the direction of the arrow returning to the sea and its color. As shown in Figure 5, it can be seen that there is a rip current phenomenon on January 2, 2015, at 3:00 p.m. on Klayar Beach, East Java, which caused 2 tourists to die. From the results of this modeling, it is found that the rip current speed ranges from 0.16 to 0.18 m/s. It is known that the rip current speed can already harm someone if it is in the range of 0.15-0.3 m/s [13].Thus, this model proves that Hydrodynamics Models can validate the rip current related to this research.

4. Conclusions

According to the results of this study, it can be concluded that hydrodynamic modeling with Delft Dashboard can be used to determine the occurrence of the rip current. This can be proven by the results of the validation which are relatively the same starting from high of water level results with RMSE of 0.16 meters and wave heights ranging from 3.4 meters. From the results of this modeling, it is found that the rip current speed ranges from 0.16 to 0.18 m/s, where it is known that the rip current speed can already harm someone if it is in the range of 0.15-0.3 m/s. These results also prove that natural occurrences in the form of rip current occurring on Klayar Beach can be proven scientifically. It is hoped that by socialization in focus group discussion these outcomes can be used by parties involved in making policies for the development of coastal tourism in the southern region of the sea of East Java. Due to the outcomes of the research still use a coarse resolution grid, thus producing better results of the hydrodynamic model in future research, it is needs a powerful computer/server due to computing and analyzing data.

Acknowledgement

The authors would like to express appreciation for the support of the sponsors of the Department of Engineer Profession Program Petra Christian University, PCU project number: 11/HBK-Penelitian/LPPM-UKP/I/2020 and Directorate General of Higher Education Indonesia and Kopertis VIII, project number: 002/SP2H/LT/K7/KM2017.

References

[1] Shalihati S F 2014*Geo Edukasi* **3** 115-126
[2] Brighton B, Sherker S, Brander R, Thompson M and Bradstreet A2013 *Nat. Hazards Earty Syst. Sci.***13** 1069-75
[3] Deskaranti R, Prasetyawan I B and Kunarso K2017*J. Oceanogr.***6** 213-220
[4] Drozdzewski D, Roberts A, Dominey-Howes D and Brander R 2015*Aust. Geogr.* **46** 15-32
[5] Rosyadi S 2018 Revolusi Industri 4.0: Peluang dan tantangan bagi Alumni Universitas Terbuka
[6] Kuswandi and Triatmadja R 2019 *J. Appl. Fluid Mech.***12**1395-406
[7] Agus I, Alwys M and Yolanda N 2020 *IOP Conf. Ser. Mater. Sci. Eng.* **846** 012072
[8] Dalrino A, Indra A and Syofyan E 2018*Int. J. Civ. Eng. Technol.* **9** 2054-60
[9] KuswandiKuswandi, Triatmadja R and Istiarto 2016 *Congress of the Asia Pacific Division of the International Association for Hydro Environment Engineering and Research* **20**
[10] Hermawan S 2016 *Civ. Eng. Dimens.***18**
[11] Hermawan S 2015 *Adv. Environ. Biol.* **9** 21-28
[12] Hermawan S, Purnomo J and Tandra D 2018 *Int. J. Civ. Eng. Technol.***9** 453-460
[13] Moulton M, Dusek G, Elgar S and Raubenheimer B 2017*Weather Forecast.* **32** 1659-66

AUTHOR INDEX